Ultimate Beekeeping Guide

Build, Manage, and Profit From Your Own Bee Colony

Owen Hayes

© Copyright 2024. All Rights Reserved.

This document is geared towards providing exact and reliable information concerning the topic and issue covered. The publication is sold with the idea that the publisher is not required to render accounting, officially permitted, or otherwise qualified services. If advice is necessary, legal or professional, a practiced individual in the profession should be ordered.

In no way is it legal to reproduce, duplicate, or transmit any part of this document in either electronic means or printed format. Recording of this publication is strictly prohibited, and any storage of this document is not allowed unless with written permission from the publisher. All rights reserved.

The information provided herein is stated to be truthful and consistent. Any liability, in terms of inattention or otherwise, by any usage or abuse of any policies, processes, or directions contained within is the sole and utter responsibility of the recipient reader. Under no circumstances will any legal responsibility or blame be held against the publisher for any reparation, damages, or monetary loss due to the information herein, either directly or indirectly.

Respective authors own all copyrights not held by the publisher.

The information herein is offered solely for informational purposes and is universal as such. The presentation of the information is without a contract or any guaranteed assurance.

The trademarks that are used are without any consent, and the publication of the trademark is without permission or backing by the trademark owner. All trademarks and brands within this book are for clarifying purposes only, are owned by the owners themselves, and are not affiliated with this document.

Table Of Contents

Part 1: Introduction 6

    Chapter 1: Beekeeping and its Importance   7

Part 2: Getting Started..16

    Chapter 2: Understanding Beekeeping   17

    Chapter 3: Choosing Your Bees   36

Part 3: Beehive Construction 52

    Chapter 4: Types of Beehives   53

    Chapter 5: Building Your Own Beehive   57

Part 4: Colony Care and Management 70

    Chapter 6: Installing Your Bee Colony   71

    Chapter 7: Seasonal Hive Management   84

    Chapter 8: Hive Inspection and Maintenance   90

Part 5: Sweet Harvests..97

    Chapter 9: Honey Harvesting   98

    Chapter 10: Working with Beeswax   106

Part 6: Transforming Hobby into Hustle................................111

    Chapter 11: Selling Your Honey and Beeswax   112

    Chapter 12: Expanding Your Beekeeping Operation   125

Part 7: Tips and Tricks 129
    Chapter 13: Insider Tips for Successful Beekeeping   130
    Chapter 14: Troubleshooting Common Problems   136

Conclusion 147
    Reflecting on Our Beekeeping Journey   147
    The Future of Beekeeping   150

Bonus Section 153
    Beekeeping Calendar   153
    Hive Inspection Checklist   156

References i

Part 1: Introduction

Chapter 1:
Beekeeping and its Importance

What is Beekeeping

Beekeeping refers to maintaining the colonies of bees in artificial beehives. The scientific name for rearing honeybees is apiculture. You breed and raise queen bees and other worker bees and ensure that their colonies are healthy.

The *Apis* genus of honeybees is the most commonly bred artificially. Apis mellifera, known as the Western honey bee, is especially highly valued for its honey production and pollination efficiency. Another species in this genus, *Apis Cerana*, or the Asian honey bee, is commonly kept in various parts of Asia due to its adaptability to local conditions.

Besides these, stingless bees of the *Melipona* genus, such as *Melipona Beecheii*, are also maintained for their high-quality honey. Additionally, *Apis Dorsata*, the Giant honeybee, is kept for its significant honey yield in South and Southeast Asia.

The *Bombus* genus, particularly bumblebees like *Bombus Terrestris*, are bred primarily for their pollination abilities in greenhouse environments. Each of these species brings unique benefits to beekeeping and agriculture.

Commercially, we breed bees in apiaries, which are large areas with many beehives. These are usually set in locations with sufficient bee pastures that provide bees with a lot of flowers as their food. Beekeeping is a popular hobby. It's widely practiced in large cities, villages, farms, and rangelands. From the Arctic to the Equator, people create bee colonies, breed honey bees, and get fresh honey from them.

Besides being a popular hobby, it's gratifying, too. For one reason, the obvious one is that you get natural and fresh honey straight out of your beehive. You can consume it yourself or sell it for money. Besides honey, you also get other valuable products like beeswax and royal jelly.

Beeswax can be used to create candle wax. It's also a commonly used product in the cosmetics industry. On the other hand, the royal jelly of your bee colony can be used as a dietary supplement. So honey is not the only benefit you'll get from this hobby.

But wait a minute. Beekeeping is not all roses. It's the job of a responsible person. You cannot start beekeeping with a lot of zeal and zest to give up a few months down the line, forgetting about your bee colony dying in your backyard. It requires discipline and commitment. Only then will you be able to reap the benefits.

The structure of a bee colony is something like this: There are around 10,000 to 60,000 bees in a colony, and they have a strict division of labor among them. The job of the queen bee is to lay thousands of eggs. The male bees are called drones, and they fertilize these eggs.

When these eggs hatch, larvae emerge. These larvae are fed royal jelly until they mature into a female bee, who lays eggs, or a male one, who fertilizes the eggs. All the female bees that are not queen bees are called worker bees; their job is to collect nectar from the flowers.

Importance of Bees in Our Ecosystem

Bees do a pretty important job in our environment. Some aspects we just discussed are producing honey, beeswax, and royal jelly. Honey has a very high nutritional value and contains other antioxidants and antibacterial ingredients. Consuming honey helps improve blood sugar levels, improve heart health, heal wounds, and suppress cough, among many other benefits. Making and harvesting honey is also advantageous to beekeepers, as is homebrewing or farming.

Other products we get from beekeeping include propolis and Bee pollen. Propolis, a substance collected by bees from plant sources, has been reported to have several health benefits, such as enhancing immune response and skin repair. Beekeepers can use the propolis bees collected for consumption or sale.

Bee pollen is a superfood harvested from hives that is packed with nutrients. It has over 250 bioactive compounds. Bee pollen is normally marketed as an antioxidant, anti-inflammatory, or dietary supplement that has positive impacts on consumer health.

Another significant advantage these bees grant our ecosystem is that they are primary pollinators. This is how it works. Many plants depend on bees for their pollination. For these plants to produce their next generation, they require their pollen grains (which include their sperms) to transfer from their male flower plants to female flower parts, where they fertilize the egg cells and produce seeds.

This transfer of pollen grains is called pollination. Some plants have pollen grains that are easily dispersed by wind, while a lot of other nuts, fruits, vegetables, and berries plants need these bees to transfer their pollen grains when they suck nectar from flowers. Otherwise, they won't be able to reproduce.

In fact, according to the Food and Drug Administration, about one-third of American crops are pollinated by honey bees. Imagine a world without almonds, cashews, sunflowers, sesame, apples, apricots, blueberries, and broccoli. Without bees pollinating these plants, we wouldn't be enjoying their fruits. Due to these very reasons, bees contribute enormously to a country's economy. In the United States alone, honeybees were worth $321 million in 2021 [1]

Beekeeping is also a very aesthetic experience. It's an absolute delight to see how bees work. Keeping bees adds to the beauty of nature. Other than the benefits of food and money, it has been found that the venom of honeybees contains toxins that are effective against human immunodeficiency virus (HIV).[2] So, these little creatures might be helping us fight AIDS in the future.

Beekeeping has many advantages, such as honey production, the preservation of the environment, economic benefits, health supplements, and pleasure. It is a fun and fulfilling activity that has the potential to alleviate stress, foster friendships, and promote the consumption of locally-grown produce.

How to Use This Guide

This is the Ultimate Beekeeping Guide, which will teach you all you need to know about beekeeping. This guide is for every beekeeper, be it a beginner or an expert who wants to add to their knowledge in beekeeping. The organization of this book is easy to follow and as informative as possible so that you can start beekeeping with enthusiasm and confidence.

When you start reading, you will realize that the book is sectioned into various parts, each dealing with different aspects of beekeeping. It has been deliberately arranged to guide you from the rudimentary to the professional level. In the introduction, you will learn what beekeeping means, the importance of bees in our environment, and how this guide will be helpful to you in your bee-keeping activities.

The guide's structure is simple to follow. The Table of Contents at the start of the book serves as a guide that shows all the chapters and sections. This makes it possible for readers to look for particular topics or go to sections they wish to repeat. The topics are arranged sequentially, and therefore, if you are a first-time beekeeper, it is recommended that you read the book from cover to cover. But if you want specific knowledge of a particular aspect like honey collection or seasonal management of hives, you can directly go to that part.

For those just starting, Part 2 is the section that will help you. Here, you will be able to read the brief history of beekeeping, its advantages and definitions, and how to select bees. This part creates the background and arms you with the information that you will require as you venture into beekeeping. Realizing the various types of bees and their traits, identifying where to acquire the primary colony, and knowing the legalities are the initial procedures in establishing the apiary.

Once you have your bees, Chapter 4 will help you understand the basic types of beehives, including Langstroth beehives, Top Bar beehives, and Warre beehives. Here, you will find complete information on how to construct your beehive, what tools and materials are necessary, as well as detailed guides for each stage of the construction and recommendations on how to prepare the hive for the bees. You may opt to buy a ready-made hive, but knowing the basic building procedures will be helpful when maintaining and modifying it.

Chapter 5 is where the actual beekeeping process is effectively implemented. This part deals with all aspects of setting up your bee colony and how to handle the colony throughout the seasons. It will help you know how to introduce bees to the hive, essential management, and deal with some fundamental problems.

The colonies are managed according to the seasons, so the right time is known to perform the right task on the hives throughout the year. Biweekly inspections are essential, and this section provides methods and recommendations to solve problems before they escalate.

The joys of beekeeping culminate in chapter 6, wherein you will learn the suitable approaches to honey harvesting, honey extraction, and processing of honey, as well as honey storage and packaging. Furthermore, this section focuses on beeswax management, including harvesting beeswax and producing different goods. If you want to consume your honey and beeswax or give them to other people, these chapters contain all the information that will help you get the maximum out of the yields.

For those considering turning their beekeeping hobby into a business, Part 6 is inevitable. This part discusses the marketing approaches, establishment of an online store, and farmers' markets. You will also learn how to grow the beekeeping business, install a new hive and other beekeeping techniques, and diversify the products. This part of the guide is intended to assist you in expanding your beekeeping options and achieving the greatest return on investment.

Even seasoned beekeepers can benefit from Chapter 8. Here, you will learn bee-keeping secrets, bee health, bee pests and diseases, and how to increase honey production. It also focuses on creating a beekeeping community, as it is always helpful to have friends who will support you and share valuable information with you. This part also includes information about the identification of diseases and pests, the ways of treatment, and prevention of these problems.

We want to remind the reader that beekeeping is as much a science as it is an art, as you will read further on. Most of the tips are general, but each apiary is different, and you may have to tweak some of them. It is recommended that while reading, you make annotations of questions, personal observations, or ideas on a separate sheet of paper. Writing a journal will assist in noting the learning process and the experiences in the course so that the information can be referred to in case of any complications.

We also have a surprise for you – the bonus section at the end of the book. In this section, you will have a Beekeeping Calendar that will assist you in organizing yourself about the seasons in the beekeeping process. Furthermore, there is a Hive Inspection Checklist, a functional checklist that will help you to perform a hive inspection properly. These tools are intended to help you as you go through the main parts of the book and as you take care of your apiary.

In the last chapter, we discuss the beekeeping experience and the prospects of beekeeping. This section is to motivate you and remind you of the massive difference you are making by enhancing the health of our ecosystem and the health of our planet. Beekeeping is not only a pastime or an investment; conversely, it is a duty that one has towards the environment.

Welcome again to the Ultimate Beekeeping Guide. This is our hope, and we hope it will be a guide that you will consult regularly. Have fun, and keep learning about the things you're passionate about.

Part 2: Getting Started

Chapter 2:
Understanding Beekeeping

History of Beekeeping

Humans have been keeping bees for centuries. Over 10,000 years ago, humans started to colonize wild bees in artificial hives. These artificial hives were made of hollow logs, wooden boxes, and straw baskets that were known as skeps. In North Africa, beekeeping was done in pottery vessels around 9000 years ago. Similar evidence of beekeeping has also been found in the Middle East, dating back to around 7000 BCE.

Egyptians also started domesticating bees around 4,500 years ago. Their artwork shows that they used simple hives and stored honey in jars. Some such jars have also been found in the tombs of pharaohs. During the 18th century, Europeans understood the biology of bees and took beekeeping to the next level. They invented a moveable comb hive, which allowed them to extract honey without jeopardizing the whole colony.

In ancient Greece, the Crete and Mycenae regions had a pretty developed apiculture system. Beekeepers used different types of hives, smoking pots, honey extractors, and other equipment. Beekeeping was considered a high-profile industry even back then, and people were dedicated to overseeing the bees. We also find Aristotle discussing beekeeping and the lives of bees.

The ancient Chinese also practiced beekeeping, primarily using wooden boxes for this purpose. Evidence of this practice has also been found in the Maya civilization. They kept and bred a species of stingless bee and used it for various purposes, including making mead-like alcohol drinks. They were the founders of the highest levels of stingless beekeeping practices.

The evidence from these ancient civilizations shows that beekeeping isn't something humans have started doing recently. It's a centuries-old practice, and humanity has always advanced it and reaped its benefits.

Basic Terminology

This is your last station before you set off on your journey of beekeeping. In this extensive list, we have put together all the vocabulary and terminology you need to know for your beekeeping career. You don't need to remember all of it now; read it, and you'll get accustomed to these words as you move further along this book.

Abdomen: The third segment of the body of a bee that contains the honey stomach, actual stomach, intestine, sting apparatus, and genitalia.

Absconding swarm: A whole group of bees that leaves the hive due to disease, wax moths, heat or water, lack of food, or other causes.

Afterswarm: A small swarm that may emerge from the hive after the first or the primary swarm has emerged. These afterswarms are generally not as large as the primary swarm and are likely to have fewer bees attached to them.

Apiary: This is a collection of colonies, hives, and other equipment used in beekeeping that are located in one place for ease of management, also called a bee yard.

Apiculture: The rearing of bees for honey production.

Apis mellifera: This is the scientific name of the western honey bee, which is native to Europe and Africa but is now found in many parts of the world.

Bait hive: A hive or a box, often located on a higher level, which is intended to attract swarms and, ideally, capture them.

Bee blower: A motorized blower that is one of the techniques for evacuating bees from honeycombs. Normally, frames are not pulled out of supers before using the blower.

Bee bread: A fermented food product made from the pollen collected by bees and mixed with nectar or honey stored in the cells of a comb. Pollen is the main type of pollen that bees collect, and it is used mostly by nurse bees to make royal jelly, which is used to feed the young larvae.

Bee brush: A small brush used to remove bees from combs without hurting them. Sometimes, it is more disturbing to the bees than simply shaking them off the frame.

Bee escape: A tool that is employed to evacuate bees from honey supers or structures as it allows the bees to go in but not out.

Beehive: A man-made structure in which bees live and reproduce, typically a box or series of boxes with removable frames.

Bee metamorphosis: The different stages that a bee passes through as she matures. The life cycle of a fly is composed of four stages, which include the egg stage, the larval stage, the pupal stage, and the adult stage. In the pupal stage, which takes place in a capped cell, large nutrient reserves are utilized to metamorphose the internal and external structures of the bee.

Bee space: 3/8-inch gap between combs and hive components or other hive parts where bees do not build, comb, or store only a small amount of propolis. Bee spaces are used as passageways to navigate within the hive. This is the space that the bees create between and around the combs in nature, which should not be violated when designing hives.

Beeswax: A combination of organic acids, fatty acids, hydrocarbons, and other substances produced from four pairs of glands on the worker bee's abdomen and used to construct a comb. Its melting point is from 143°C to 147°C, and its boiling point is from 200°C to 215°C—6 to 147.2°F.

Bee venom: The poison produced by the glands located at the base of the stinger of the female bee.

Boardman feeder: A feeder used to feed bees. It is made of an inverted syrup jar with a base that fits into the hive's entrance.

Bottling: Filling bottles or jars with honey, either by gravity, where the honey flows into the jars, or a honey bottling machine, where the honey is pumped into the jars faster.

Bottom board: The floor of a hive where all the other parts of the hive are placed on top of it. It can be a solid bottom board, a screened bottom board that lets the debris on the hive fall through the ground, or a waiting catch board.

Brace comb: A small amount of stabilizing wax is placed between two combs or frames to secure them. A brace comb is also constructed between the comb and other hive components or between two hive components, such as the top bars.

Brood: The young bees that are still in their cells. Brood can be in the form of eggs, larvae, or even pupae of various stages of development.

Brood chamber or Brood nest: The chamber where the queen rears her young; it may consist of one or more hive bodies and combs.

Burr comb: A small wax formation on a comb or a wooden appendage in a hive but not attached to any other part.

Capped brood: Pupae whose cells are closed with a porous lid by the adult workers to prevent the young ones from feeding during their nonfeeding stage of development.

Cappings: A thin layer of wax covers the total number of honey cells. This layer of wax is usually cut from the surface of the comb with honey so that honey can be harvested.

Caste: A term used to refer to the subgroups of the same species and sex of social insects that have different body structures or behaviors. There are two classes of bees in honey bees, namely the workers and the queen bees. The drones are of a different sex and, therefore, not a caste, even though they are often misclassified as the third caste.

Cell: The hexagonal chamber constructed by honey bees to store honey and bee larvae.

Chilled brood: These are bee larvae and pupae killed by cold temperatures. They usually occur during spring, when the colony's size increases rapidly, and at night, when few bees are present to warm the brood.

Chunk honey: Honeycomb cut from frames and then packed with liquid honey in jars.

Clarifying Tank or Clarifier: Any tank or holding vessel that stores honey briefly while the wax and other materials float on the surface and separate from the honey.

Cluster: A large group of bees clinging to one another, one bee on top of another.

Colony: The queen, her offspring, the drones, and all the workers residing in one or the other nest, hive, etc.

Comb: A sheet of six-sided wax cells constructed by honey bees to contain brood, honey, nectar, and pollen. A sheet of comb will have a layer of cells on each side of the comb, and these cells will be joined at their base.

Comb honey: Honey that is sold still in the comb. It can be made by slicing the comb from the frame or by constructing the comb in unique frames that enable the honey to be easily extracted without the need to cut the comb.

Creamed honey: Honey that has been allowed to crystallize under controlled conditions so that it forms tiny crystals and has a very smooth texture. An initial charge or seed is sometimes employed to regulate the crystallization rate.

Crimped-wire foundation: A type of foundation in which crimped wire is placed vertically during production. The wire also adds to the strength of the foundation.

Crystallization: The process of sugar forming crystals in honey.

Dance language: The system of signals that honeybees use to describe the positions of food sources or other sites that may be suitable for establishing a nest.

Deformed wing virus (DWV): A honeybee virus that spreads through bees and the feeding of varroa mites.

Division board feeder: This is a wooden or plastic box hung in a hive in place of one or more frames and contains feed for bees.

Double screen: A thin wooden frame with two layers of screen. It divides two colonies into a single hive, one on top of the other. Entrances are made on the upper side to enable the upper colony to access the outside world.

Drawn combs: Frames with the foundation the honeybees have drawn out.

Drift: The inability of bees to return to their hive in an apiary that contains many hives because they have entered other hives.

Drone: Male honeybees that grow from unfertilized eggs.

Drone comb: A comb measuring about four cells per linear inch used by bees for the rearing of drones and for storing honey.

Drone layer: A queen with no sperm in her body and thus cannot lay viable eggs. Consequently, all the young she lays will be drones.

Dysentery: A disease of adult bees manifested by diarrhea and may be attributed to starvation or poor quality of food, overcrowding due to adverse weather conditions, or nosema disease.

European foulbrood: An illness that affects only the brood of honeybees and is bacterial.

Extracted honey: Honey in liquid form, which has been taken out of the comb.

Extraction: The final step of honey production is where honey is separated from the combs using a honey extractor through the use of centrifugal force.

Extractor: A machine that removes honey from the cells of a comb by using force that tends to throw objects away from the center of the circle.

Fermentation: The process where yeast feeds on sugar and synthesizes alcohol during digestion. Bees evaporate water as nectar is converted into honey before fermentation can take place.

Fertile queen: A queen with enough sperm stored from the drones to lay fertilized eggs.

Follower board: A narrow board the size of a frame that is placed into a hive to take up space and limit the bees. Sometimes, it is done to assist the smaller colonies, which may struggle to warm the brood nest adequately.

Foragers or Field bees: These are worker bees, which are at least two to three weeks of age, and their main duties are to forage for nectar, pollen, water, and propolis.

Foundation: An artificial thin sheet of beeswax or plastic that has the shape of cell bases embossed on both sides. Applied to promote straight combs in frames and to provide additional support to the combs constructed on the base.

Frame: A wooden or plastic body, rectangular in shape, on which a comb is built and which is inserted in a hive box.

Fructose: This is the most common monosaccharide in honey.

Fume board: A rectangular cover of absorbent material on the underside. A chemical is applied to the material to harvest the bees from the supers for honey harvesting.

Glucose: One of the two principal sugars in honey; it crystallizes during crystallization or granulation. Also called dextrose.

Grafting: Transferring a young worker larva from its cell to a queen cup to rear it into a queen.

Grafting tool: A probe, needle, or scoop moving larvae from worker cells to queen cells.

Hive: An artificial structure created by a colony of bees to use as a nest chamber.

Hive body: A box containing frames of comb. It is used explicitly for the brood chamber to differentiate it from the supers, but it can be used for any hive box.

Hive stand: This structure supports the hive and raises it slightly off the ground.

Hive tool: A tool made of metal used to open hives, pry and handle frames, and scrape wax and propolis from the equipment.

Honey: A natural sweet substance made by bees from the nectar of plants and stored in their combs. It is made up of a solution of sugars and approximately 17% water. It also has low proportions of mineral matter, vitamins, proteins, and enzymes.

Honeydew: Honeydew is a sugary substance secreted by aphids, leafhoppers, and some scale insects collected by bees, especially when there is no nectar. It can be used to produce what is commonly called "forest honey" or "honeydew honey."

Honey House: A structure where honey is collected from combs and where the honey collection equipment is kept.

Honey stomach or Honey crop: A sac-like structure in the honeybee's digestive tract that can be distended and used for carrying nectar, honey, or water.

Inner cover: A thin cover placed on the top of the beehive below the standard telescoping outer cover.

Invertase: A substance secreted by honey bees and used to convert sucrose into glucose (dextrose) and fructose (levulose).

Larva (plural, larvae): The second developmental phase of bees. A white, legless worm: like an insect.

Laying worker: A worker that lays eggs that do not contain sperm and, therefore, only produces drones, typically in colonies that are irretrievably queen-less.

Mating flight or Nuptial flight: It is a flight made by a virgin queen in mid-air to match multiple drones.

Mead: Honey wine.

Migratory beekeeping: The practice of transferring colonies of bees from one area to another within a single season to benefit from several honey sources and to meet pollination services demands.

Nectar: A sugary substance produced in the nectary of flowering plants to entice animals to help pollinate. Nectar is the raw material used for making.

Nectar dearth: A time when flowers produce little or no nectar, which is the food that bees collect for their hive.

Nectar flow: When nectar is abundant, bees make and store abundant honey. Sometimes, it is associated with a particular flower only, such as "the goldenrod flow."

Nectar guide: Streaks or spots on flowers that are thought to lead insects to nectar.

Nosema disease or Nosemosis: A disease of honey bees caused by protozoa *Nosema Apis* or *Nosema Ceranae*, which infect the bees' intestines. The microbes kill the bee's gut, and acute infections cause diarrhea and malnutrition in the bees.

Infections of the two types have somewhat different manifestations.

Nucleus or Nuc: This is a small colony of bees made of fewer frames than a standard colony and is usually reared in a small-sized hive body. A nuc is generally made of three to six frames of comb and is used mainly in creating new colonies, queen rearing, or queen storage.

Nurse bees: These are young bees between three and ten days old. Their primary duties include feeding the developing brood.

Observation hive: A hive that is mainly made of glass or clear plastic to enable the monitoring of bees.

Pheromones: These are chemical signals which are released from glands and are used in communication.

Play flights or Orientation flights: These are short flights made in front of or close to the hive to familiarize the young bees with their environment.

Pollen: The male gametes or the microspores developed in the anthers of flowers. A food source for honeybees, which they collect and utilize in their diet to obtain protein.

Pollen substitute: Any substance like soybean flour, powdered skimmed milk, brewer's yeast, or a combination of the above used instead of pollen to provide protein to stimulate brood rearing. Usually given to a hive in early spring to promote the establishment of a new colony.

Pollen trap: A gadget used to extract pollen loads from pollen combs of arriving bees.

Propolis: Plant saps and resins collected by bees and used to strengthen the comb and seal cracks. Possesses antimicrobial and waterproofing properties.

Pupa (plural Pupae): The third stage of the honeybee life cycle, at which the larva transforms or undergoes metamorphosis to become an adult bee.

Queen: A small cage where a queen and three to five worker bee attendants are kept for transportation and release into a new colony.

Queen cage: A specially designed elongated cell in which a queen is to be bred. It is an inch or longer and dangles vertically from the comb in front of the bird.

Queen cell: A specially designed elongated cell in which a queen is to be bred. It is an inch or longer and dangles vertically from the comb in front of the bird.

Queen clipping: The process of cutting part of one or both of the front wings of a queen to immobilize her and stop her from flying.

Queen cup: A small wax cup constructed by the bees to enable the creation of new queens. After the queen has laid an egg into it, it becomes known as the queen cell.

Queen excluder: A flat sheet of metal or thin plastic with openings that allow only the worker bees but cannot allow the queen and the drones or restrict their access to specific parts of the colony.

Robbing: Robbing is a process where bees take nectar or honey from other colonies that are not their own.

Robbing screen or Robber screen: A device placed over the entrance of a hive that confuses the robber bees while allowing the resident bees to move in and out.

Sacbrood: A viral disease that infects honeybee's broods or young ones.

Scout bees: Bees that are in search of a new source of food, especially pollen and nectar, water, propolis, or a new nest for the colony that the scout bee is part of.

Secondary swarm: A minor swarm involving a new virgin queen may emerge after the primary swarm leaves.

Skep: A bell-shaped straw beehive without the use of movable frames.

Slatted rack: A wooden structure placed below the bottom board and above the first brood box. It may help minimize traffic at the hive's entrance and assist the bees in regulating the climate at the bottom of the bottom frames.

Solar wax melter: A box-like container with a transparent lid that melts wax from combs and cappings using heat from the sun.

Splitting: Reproducing a new bee colony from an existing one by breaking it into two or more colonies.

Sucrose: The most common sugar found in most nectar.

Super: Any hive body, usually a smaller box containing honey that the beekeeper wants to collect. Usually, it is located above one or several brood chambers. Supers are generally boxes of a medium or shallow size.

Supersedure: Supersedure is raising a new mother queen within the same hive to replace the old one naturally.

Swarming: Swarming is the process by which bees emerge from a hive, clinging to a tree branch or other surface, and then search for a new home.

Swarm cell: Queen cells, usually located at the bottom of the combs before swarming, are developed to supply a new queen for the colony to which the swarm originates.

Terramycin: An antibiotic used to treat American and European foulbrood. It is only available in the United States as a prescription drug under the direction of a veterinarian.

Tracheal mites: A parasitic mite belonging to the family present in adult bees' internal trachea (breathing tubes).

Uncapping: Uncapping is removing a thin layer of beeswax that covers capped honey to enable honey extraction.

Uncapping knife: A knife utilized to cut or peel the cappings off combs containing sealed honey before extraction. Some are heated by steam, while others are heated by electricity.

Uniting: Uniting is joining two or more colonies to form one big colony. The weak colonies are expected to be transferred to a single location in the fall so they can winter there.

Veil: A hat, helmet, or headpiece with a wire or fabric netting screen to shield the beekeeper from stings on the head and neck.

Virgin queen: A queen that has not yet been involved in mating.

Wax glands: Organs that produce beeswax. It is located in two on worker bees' ventral surface of the four last abdominal segments.

Wax moth: Typically, larvae of the greater wax moth can destroy brood and empty combs.

Winter cluster: A compact mass of adult bees within the hive. During winter, bees feed on honey and generate heat using flight muscles.

Worker bee: A female bee that cannot reproduce. Most of the honeybees in a colony are worker bees, and their duties are to forage for food, build and maintain the hive, protect the hive, and take care of the young ones, except for the queen bees.

Worker comb: Cell measuring approximately five cells to the inch in which workers are bred and in which honey and pollen are stored.

Chapter 3:
Choosing Your Bees

Types of Bees and Their Characteristics

Let's proceed with our beekeeping journey. Before we tell you how to start beekeeping, let's first discuss the different types of bees and their attributes. This won't only let you better decide which types of bees you want to keep but also help you become a subject matter expert on beekeeping.

Honeybees

Honeybees, the most common of them all, are social insects that live in colonies and hives. You can also find them nesting in rock cavities or hollow trees. They have their unique yellow-orange and black stripes. Also, they have a lot of hair on their body that gives them a fuzzy appearance.

The colony structure of honeybees is something you're already familiar with. The most numerous types of bees in a colony are worker bees. The job of these bees is to forage for nectar and pollen. They are also responsible for caring for the young ones. Repairing the hive is another thing that falls under the job role of these hard workers.

Queen bees like to relax and lay eggs. The reproductivity of the hives and, in turn, the colonies are ultimately dependent on the queen bees. Male bees are called drones, and they have to mate with the queen bees and fertilize her eggs. Drones are developed from unfertilized eggs.

Honeybees are the most common type of bee that beekeepers like to keep and breed. They produce honey, so they are the most beneficial. Honeybees are absolutely essential for the pollination and reproduction of a large number of plants, including sunflowers, clovers, and apples.

To keep honeybees, you must inspect the colony's health regularly. Check for diseases or pests. And take prompt action if you find any. You'll have to ensure that these bees have enough space and food. After establishing a thriving colony, you can harvest honey and other products like beeswax and royal jelly. But be careful while doing so. These bees sting when they feel a threat.

Bumblebees

Bumblebees are large, fuzzy, black, and yellow or black and orange. Like honeybees, they are social, but their colonies are much smaller. Bumblebees can fly at lower temperatures and less favorable conditions than honeybees.

Beekeepers usually don't keep bumblebees. They produce a minimal amount of honey that's only sufficient for their colony. So, you aren't getting any. However, these bees play a vital role in pollination. Tomatoes, peppers, berries, and a lot of other plants depend, to a large extent, on bumblebees for their reproduction.

Bumblebees are also crucial in the pollination of wildflowers. You can also use them to pollinate crops in your greenhouse if you're a farmer. Bumblebees can also sting when they are threatened.

Other Kinds of Bees

There are many other kinds of bees besides honeybees and bumblebees. Beekeepers don't usually domesticate these bees, but they are highly important in pollinating various crops. Let's discuss them briefly.

Longhorn Bee

Longhorn bees have hairy bodies. They flaunt black and white stripes and long antennae. You can usually find them in pre-existing cavities, hollow stems, or beetle burrows. Longhorn bees are essential for pollinating aster, sunflowers, and various legumes. They won't harm you unless you trigger them.

Green Metallic Sweat Bee

These tiny creatures have beautiful iridescent bodies that give out stunning green, blue, and gold hues. Their favorite spots to reside are rotten wood, well-drained soil, or underground tunnels. They pollinate many flowers, including goldenrod and clover. They are nonaggressive in general but can sting.

Leafcutter Bee

Leafcutter bees have slender bodies. Black and pale yellow stripes make these insects a delight to look at. They have large jaw bones and are adept at cutting leaves. Thus, the name leafcutter bees. Plants pollinated by these insects include alfalfa and tomatoes.

Carpenter Bee

These are large and black. They have a fuzzy appearance and smooth abdomen. They mostly live in caves in the woods. They are essential for the pollination of many fruit trees. They can sting but aren't typically aggressive.

Sweat Bee

Sweat bees have varied appearances. Some have metallic green bodies, some have blue bodies, and others have black and white stripes. They inhabit ground burrows and sandy areas. They pollinate alfalfa and a range of wildflowers. They are also nonaggressive but can sting when agitated.

Miner Bee

Their bodies are fuzzy with creamy yellow hairs on the thorax. They have a black abdomen. They nest in the ground burrows and pollinate plants such as irises, roses, persimmons, and parsnips. They won't sting you unless you directly handle them.

Squash Bee

These bees, just like many others we discussed, have a hairy thorax and a smooth abdomen. You can commonly find them in ground burrows near squash and pumpkin plants. They serve as important pollinators for these plants. They'll sting only when you handle them directly.

Unequal Cellophane Bee

These bees are known for their black and tan coloring, fuzzy head, and translucent wings. They nest in sandy or well-drained soils. Wildflowers such as asters and goldenrods depend on these bees for the dispersal of their pollen. They are not aggressive in general.

Ashy Mining Bee

This bee is black. It has two stripes of whitish hair, and its abdomen is black. You can find it resting in ground burrows of sandy and loamy soil. It's also a vital pollinator for plants such as primrose and forget-me-nots. They also sting, but only when handled directly.

Ivy Bee

This bee is dark brown. It has dense golden brown and orange-brown hairs. They make their nests in burrows in sandy soils. They have the name ivy bee because they are crucial pollinators for ivy flowers. They also only sting when handled directly.

Bellflower Resin Bee

The bellflower resin bee is hairy and has black and white stripes, which gives it an overall greyish appearance. It nests in cavities like hollow stems and holes in the wood. The bee is an effective pollinator for bellflowers and other related species and is generally nonaggressive.

Africanized Honeybee

These bees are orange-yellow with dark brown and black bands. You can find them nesting in enclosed places and artificial structures and cavities. They pollinate various crops and wildflowers. They are generally defensive and sting only when you disturb their hives.

Long-horned Bee

These bees have extra-long curved antennae. They have a thick fur. They are black, but their setae(legs) are whitish. They effectively pollinate plants such as sunflowers and legumes. These insects are also non-aggressive and won't sting you unless you give them a solid reason to do so.

Digger Bees

These bees are large and solitary. They have white hairs on their bodies, which give them an overall grayish appearance. They nest in burrows in sandy and loamy soils. These insects pollinate wildflowers and garden plants, among others. They are also nonaggressive.

Squares potted Mourning Bee

These fuzzy black bees have white spots over them. Interestingly, they reside in caves filled with other bees, such as mining bees. They have parasitic relationships with other bees, so if they invade some other bees' nests, it might affect the overall pattern of pollination in a crop. Like most of the other bees discussed, they are generally not aggressive.

Mason Bee

These bees have metallic green and blue bodies. You can find them nesting in preexisting cavities like hollow stems or holes in the wood. They pollinate early spring plants like fruit trees. They are also not aggressive in general.

Carder Bee

Carder bees have black and yellow markings resembling a yellow jacket on their bodies. They also nest in pre-existing cavities like hollow stems of plants and artificial bee hotels. They effectively pollinate plants of the aster and daisy families. They also pollinate herbs like lavender. They are also nonaggressive.

Striped Green Sweat Bee

These bees have a smooth abdomen with black and yellow stripes. Their head and thorax are metallic green. They rest in ground burrows in soil. They pollinate rhododendrons, irises, roses, and parsnips. They are also nonaggressive in general.

Hairy-Footed Flower Bee

These bees are small and dark-colored. They have reddish-orange hairs and tan stripes. You can find their nests in soft mortar, sandy cliffs, and mortars of old walls. They are effective pollinators for plants such as lavender and borage. They are nonaggressive in general.

Tawny Mining Bee

The color of Tawny Mining Bee is rusty, reddish brown, and golden brown. They nest in ground burrows of well-drained soil. These pollinated plants include fruit trees, dandelions, and flowers that bloom in early spring. They are also nonaggressive.

Yellow-faced Bees

These are small black bees with creamy-yellow patches on their faces. You can find them nesting in premade holes like those in deadwood. They effectively pollinate plants like dandelions and clover and are not aggressive in general.

Wool Carder Bee

These bees have black bodies adorned with unique white or yellow markings. They nest in plant cavities and use plant hairs for nest lining. They are effective pollinators for many plants, including different flowering plants. They can sting, but usually, they don't.

Southeastern Blueberry Bee

These bees somewhat resemble tiny bumblebees. Their stout body is covered in creamy-yellow hairs. You can find them nesting in loose sandy soil burrows. They pollinate plants like blueberries and other flowering plants. They sting only when you directly handle them.

Nomad Bees

These bees are similar in appearance to black and yellow wasps. They are not fond of building their nests. Instead, they inhabit the nests of other bees as parasites. They don't directly contribute to pollination, but they play a crucial role in maintaining other bee populations that affect pollination overall. Interestingly, they don't have any stringers.

Stingless Honeybees

These bees come in many appearances, the most common being shiny and metallic. They build their nests in tree cavities, man-made hives, hollow logs, etc. They effectively pollinate tropical fruits, herbs, and wildflowers. They don't have stringers and are hence nonaggressive.

Cuckoo Bees

These bees are wasp-like because they have slender bodies and are orange. They don't bother to build their nests. Yes, you guessed it right! They are parasites and invade the nests of other bees. They also don't contribute directly to pollination but keep the population of different bee species in check. They also don't have stringers.

Orchid Bees

These bees have metallic green or other shiny colors. They are hairless and nest in hollow tree cavities. They pollinate different types of orchid plants. They are nonaggressive and sting only when you handle them directly.

Sourcing Your First Bee Colony

Now that you know about every type of bee, it's time to source your first bee colony. This starts with deciding where to buy bees.

A beekeeping venture begins with finding a source for the initial bee colony. This crucial step requires selecting the appropriate method of acquiring your bees to be healthy and productive hives. It is essential to research different suppliers to find a supplier with a good reputation and healthy bees. It is wise to select local suppliers because the bees used in honey production are usually well-suited to the climate and the available flora. Check on the bees before purchasing so that you can see that they are healthy and do not have any signs of disease.

Package bees are one of the most common forms of obtaining bees by nonprofessional beekeepers. They are made up of a queen and worker bees and may contain drones, all housed in a screened box. In most regions, package bees should be ordered early in the spring to maximize the chances of the bees arriving on time for installation in the hives. New hives are best established in spring when the weather is good since this is natural for bees, and they are active at this time.

Another option is using a nucleus colony or nuc, a small functioning colony with a queen, workers, brood, and food stores in the form of frames of honey and pollen. They are tiny hives that can be moved directly to a new, more giant hive and are valuable for bee farming. They have frames, the queen, the bee colony, and some initial hive products like comb and brood. Typically populated nucs are sold for about $300 each and must be pre-ordered in winter since they are usually in high demand.

Swarming is taking a new bee colony from a parent colony that has left the original nest. If you have a swarm of bees, you can try to get them into a container using a flat object and something like a piece of cardboard. When there are many bees in the container, close the lid and take the bees to your hive. It helps to ensure that the hive is clean and, ideally, the frames the bees have to start drawing new comb on are put in place.

The other way of obtaining an entire hive with an active colony is to purchase one from another beekeeper, which is usually quite expensive. This method is beneficial because it means that the person will start with a hive that has already demonstrated success and thus does not have to do as much work to set everything up. New beekeepers usually acquire their first colony from a business selling bees or another hobbyist beekeeper. Local bee groups are also an excellent place to look for a reputable seller since most people selling bees locally are usually experienced beekeepers.

Legal Considerations and Permits

Urban residents face legal issues and the need to acquire permits, which are essential to consider before undertaking bee farming. While many restrictions on beekeeping are now less common, it is always wise to consult with your local government administration or council before you start. In many areas, you must be licensed to perform legal beekeeping.

The law regulating beekeeping plays a vital role in protecting the beekeeper and the population in general. These rules protect bees and the public and avoid any possible confrontation. The above regulations are crucial in proper and sustainable bee farming.

In the United States, beekeeping regulations vary by state and local jurisdiction. Beekeepers typically need to register their hives with the state's agricultural department. For instance, in South Carolina, beekeepers must register their hives with the Clemson University Cooperative Extension Service.[3]. Regulations include maintaining appropriate distances between hives and property lines and managing bee health through regular inspections. In Michigan, beekeepers must follow state regulations that involve registration, disease control measures, and proper hive management.[4].

In the United Kingdom, beekeeping is regulated under the Bee Diseases and Pests Control Regulations 2006. This legislation provides specific measures beekeepers must take to prevent and control bee diseases like European foulbrood and American foulbrood. These regulations are enforceable by the National Bee Unit, part of the Animal and Plant Health Agency. Beekeepers are required to register with the BeeBase database to help manage disease outbreaks and facilitate easy communication with government inspectors[5].

In Canada, beekeeping regulations are managed at the provincial level, with each province having specific guidelines that must be followed. For instance, in British Columbia, the Bee Regulation under the Animal Health Act requires all beekeepers to register their hives with the Ministry of Agriculture[6]. This regulation includes detailed requirements for disease management and hive placement to ensure bee health and public safety.

In many places, beekeepers are legally allowed or required to get permits or licenses for beekeeping. These permits, often obtained from the local agricultural department or zoning authority, let the local beekeeper demonstrate that they are knowledgeable and responsible about beekeeping. Suppose you are applying for a permit or license for beekeeping. In that case, you may be required to give details of your experience in beekeeping, the number of beehives you intend to keep, and where the apiary is located.

Certain areas may also require the beekeeper to attend training or education classes to be willing to practice proper beekeeping. Acquiring the permit and license meets the legal requirements and shows commitment to bee farming and the bees. Such regulations help to balance the beekeepers' and the community's interests and the general health of the bees.

There are specific regulations of beekeeping depending on the zip code in which the beekeeper resides, and such rules are put in place for the safety of the beekeeper and the general public. Restrictions on the number of hives per farm help avoid the overpopulation of bees and balance the bees and their activities with the ecosystem. These restrictions are usually based on local laws, and the number of beehives one can keep depends on these laws.

Rules regarding minimum distances from neighbors are set to reduce issues related to stinging bees and possible disturbances of neighbors' dwellings. As a beginner, it is recommended that you consult with your neighbors before beginning beekeeping, especially if any of your neighbors have a bee sting allergy. This way, they will be relieved of their fears and misconceptions about bees and beekeeping. By following these requirements, no conflict is likely to happen, and this will foster a good relationship between community members.

Zoning restrictions of beekeeping establish areas of the locality in which beekeeping activities are permissible. Commonly, there are different regulations for different zones, as needs and issues may differ in the community. Safety is always essential in beekeeping, so regulations touch on several aspects.

Part 3:
Beehive Construction

Chapter 4:
Types of Beehives

In this chapter, we will discuss different types of beehives. There are three common types:

Langstroth Hives

Langstroth hives consist of a series of boxes known as supers, which are placed atop one another and may be added or withdrawn as needed. They are the most commonly used by commercial beekeepers because of their ease of use and ability to produce large honey yields. This type of hive is named after Lorenzo Langstroth, a reverend who, in 1851, observed that the bees always leave a space of 3/8 inch between the combs. If the space is more significant, the bees will construct combs; if the space is smaller, the bees will cover the space with propolis. Knowing this, Langstroth made a revolutionary innovation in the design of beehives. His brood box was equipped with frames hanging from the top edges of the brood box, leaving a 3/8" gap between the frames and the brood box.

Some advantages of the Langstroth hive design include: These frames are easily transferable, thus making the inspection and honey harvesting process relatively easy. This also allows beekeepers to monitor the diseases, pests, and overall hive conditions without affecting the entire colony. The ability to stack the boxes makes it possible for the beekeepers to add more supers to accommodate big honey production. Also, Langstroth-type hives are widespread and have numerous resources, equipment, and beekeepers' knowledge backing them.

However, some drawbacks have been associated with Langstroth hives. The boxes can become rather hefty when laden with honey, and some beekeepers might find lifting and carrying them hard. Also, Langstroth hives may be more costly to buy than other hives.

Top Bar Hives

Top Bar Hives have one long box used to construct bars across the top where bees create comb. For example, they are popular among amateur beekeepers and people who dislike using synthetic chemicals on bees. Also called top bar hives, they are preferred by many beekeepers due to their natural ease and the bees' tendency to build their combs without having to be guided by the keeper. To foster the formation of a wax comb, bees are encouraged to hang from a single bar of wood, which enables them to construct the wax comb as they wish. This is different from the Langstroth hive, where one uses a four-sided frame to direct the bees.

However, this method yields relatively tiny amounts of honey, although some individuals consider the clear yellow comb that is harvested superior to liquid honey. Since no additional support exists, the combs are easily removable during harvesting, and an extractor is unnecessary.

As mentioned, Top Bar hives have the following advantages. The fact that the hives are laid horizontally to form a pyramid reduces the required lifting amount, making it easier for beekeepers with some form of disability who may find it difficult to lift heavy loads. Colonies were also found healthier when bees constructed combs on the top bars. Furthermore, Top Bar hives are considerably simple and cheap to build, thus making them suitable for people interested in do-it-yourself projects.

Nevertheless, Top Bar hives have some limitations, which are characteristic of this type of hive. Bees may build combs that cross over bars, posing a challenge during inspection and management. This means the comb should be kept straight by regularly inspecting it to maintain a straight edge. Also, the Top Bar hives usually produce less honey compared to the Langstroth hives, primarily because they focus more on the ability to make enough to feed the colony rather than honey production.

Warre Hives

Warre Hives are like the top-bar hives but with the bars oriented vertically. It is familiar with hobbyist beekeepers and those who practice natural beekeeping. The orientation of the Warre beehive is such that it mimics the positions that wild bees naturally prefer. It is less expensive than the Langstroth hive and less complicated to manage than other hives; thus, it is suitable for the busy beekeeper who does not interact much with his bees.

The advantages of Warre hives include fewer manipulations of the combs and the natural development of the combs, which are favorable for bees. The hive design fits natural beekeeping principles, allowing little interference with bees' activities. Also, Warre hives can be more straightforward and less expensive to build than other hives, and thus, they can be built by anyone.

Nevertheless, some difficulties can be connected with Warre hives. Beekeeping, in this case, can be more strenuous and invasive to the bees compared to the Langstroth hives. Also, there is a limited number of books and other materials about Warre hives, and there are fewer communities for people with such hives, so it may be more difficult for novices to find help.

Chapter 5:
Building Your Own Beehive

There are three types of beehives, but you can also create your own—it's not that complicated. In this chapter, we'll discuss what materials and tools you need to construct your beehive, how to build it, and how to set it up.

Essential Materials and Tools

Materials

To construct your beehive, you must purchase several materials to create a strong and stable beehive. The most common material used in the construction of beehives is wood, owing to its ability to offer extended service without decaying.

You should use wood that does not quickly get rotten and decayed, particularly when building the hives. You can use cedar or pine wood for the construction. These types of wood are very hard-wearing and can readily withstand seasons of rain and scorching heat. Also, the wood should be fresh and not treated because chemicals used to treat such wood may harm bees.

If you want a more environmentally friendly product, look for the timber sourced from certified forests. The wooden parts should be about 3/4-inch (19mm) thick to provide insulation and support to the bees. Cut the wood into the following dimensions: The standard size of the hive boxes should be a Width of 16 1/4 inches (41cm), a Length of 19 7/8 inches (50cm), and a Depth of 9 5/8 inches (24cm). The frames of the hives should be 19 inches long and 1 3/8 inches wide or 48cm long and 3. 5 cm wide.

The nails and screws are needed to fix various parts of the beehive as they are required during the construction of the beehive. When designing a building, ensuring the right type and size of fasteners are used to guarantee a strong structure is crucial. When choosing nails and screws, go for stainless steel or galvanized to avoid the risks of rusting and degradation of metal.

It is recommended to use 1 1/4 inch (3. 2cm) stainless steel nails or screws to join the different hive boxes together. As for the fastening of the frames, 3/4 inch (1. 9cm) stainless steel nails or screws are suggested.

It is essential to put a protective coating on the outside of the hive to increase its longevity and prevent it from being broken by the weather. Bees are not affected by paint or stain, but it should be investigated to find a toxic-free paint or stain to protect them. Select a water-based paint or stain and make sure it is designed for exterior surfaces. Coat the entire exterior of the hive, the hive boxes, and the cover and base with paint or stain.

Mesh wire is used to produce a screen bottom board for the hive to provide ventilation and pest control. This is because it creates a screen through which debris and varroa mites can pass, thus minimizing infestation. Choose a high-tensile strength, abrasion-resistant mesh wire with a wire gauge of 1/8 inch (3mm). Trim the mesh wire to match the size of the hive base so that it will fit perfectly on the hive base.

Tools

If you are thinking of constructing the beehive on your own, then it is essential to note that tools are also necessary. These tools shall help you in the construction of a solid and efficient hive for bees, which will guarantee the bees a good shelter. Below are the tools you will require to construct your beehive.

Saw: You will need the saw to make several cuts, which are vital in constructing your beehive. It assists you in cutting wooden parts to the standard lengths and sizes you desire.

For this, you can use a handsaw or a circular saw, but if you have an opportunity to use power tools, a circular saw is preferred. Remember to ensure that the saw you choose is well-sharpened so that the surface finish is very smooth when you use it to cut.

Hammer: Hammer is another essential tool that must be employed in constructing a beehive during the construction period. It is utilized when fixing wooden pieces by driving the nails through the wooden surface. Speaking of hammer selection, one should pay attention to the hammer type, weight, and grip. A compact hammer that is easy to hold should be used so that you do not put much pressure on the arm's hand or muscles. Also, you will have a claw hammer, which is useful when pulling out a nail, perhaps because it is bent or sticking out.

Drill: A drill is an all-rounded tool useful when constructing bee hives, some of which are mentioned below. It is also used for making holes for screws, ventilation, or any other part that may be required in a specific design.

Two factors are important when using a drill: its capacity and size. A cordless drill can be more versatile because you can move around from one region to another without worrying about where the power outlets are.

Measuring Tape: It is essential to measure the pieces correctly so that the beehive pieces fit and interlock properly. Most of the work will involve measuring tape, so it should be your close companion. This will assist you in determining the actual sizes of the hive boxes, frames, and other needed parts. Ensure that you get the best strong measuring tape that can withstand pressure and comes in inches and centimeters.

Step-by-Step Construction Guide

This section will discuss the steps you need to take to construct your beehive. The first step to creating a beehive base is to determine the correct size of the wood pieces you will need based on the design of the specific bee hive that you have in mind.

The base can be a solid bottom board and four support legs, or it can be made of other elements, but it should be noted that the base is not always required in the design of a pallet. After that, use nails or screws to fix the support legs on the bottom board so that they will be well secured.

Re-measure the base and adjust it with greater precision to ensure it is flat and firm. After the base is built, it is recommended that you cover it with paint or stain to enhance its appearance and protect it from external factors.

Constructing the Hive Boxes

To build your hive boxes, first, you need to establish the number of bees in the colony and the size of the hive boxes you wish to construct, then cut the wood. Afterward, ensure that either nails or screws join all the pieces of the boxes to ensure that they are well-fixed.

Do this for every box, creating several layers appropriate for the bees. Finally, you may wish to include other features on the side of the box to enable easy lifting and moving of the container.

Adding Frames and Foundation

For the last step of constructing hives, properly arrange the frames inside the hives boxes and ensure they are at the proper distance. This helps the bees move within the hive and form their comb. Insert the foundation into the frame slots and align it, but with some tightness or pressure. A Beeswax foundation will be used to guide the development of the comb, while a plastic one will be used to guide comb build-up. This has to be done for all the hive boxes, frames, and foundations, which should be put where needed, and space should be placed between the frames so that the bees can move around and construct their combs appropriately.

Installing the Hive Cover

Finally, place the hive cover on top to complete the construction of your bee hive. Measure the wood and then sew it to a size slightly larger than the hive boxes to cover the boxes completely. Last but not least, fix the cover to the hives boxes either by nailing or screwing at the upper side of the cover. More ports can be created on the mesh wire if needed.

When placing the hive, ensure it is on the ground and appropriately positioned in line. If necessary, you should modify it to fit newcomers in the hive. You have now developed your bee hive from the bottom to the top without any help from anyone. It is now the right time to choose the right place to place it, ensure enough air circulation, and guarantee that the hive is safe from predators. But do not worry. The following section is about just that.

Setting Up Your Beehive

Finding the Perfect Location

To ensure that your bee colony thrives, you must choose a location with the following considerations in mind.

The ground level should be flat. It should also be dry so the hive is stable and you don't have to face mold issues. Keeping the ground dry is also essential in preventing moisture, which isn't a good thing for a growing bee colony.

Ensure that the location is near the source of food and water. Bees need flowers to make honey, so you must ensure that flowers are nearby. You can plant a small garden near the colony with different flowers blooming at other times of the year.

Also, bees need water for drinking and also for cooling their hives. So, there should be a small pond or lake near the colony. If that's not feasible, don't worry. Just place a water dish near the colony and replace the water regularly.

Another crucial aspect is the hive's positioning. Your hive should be positioned so that it receives sunlight in the morning and shade in the afternoon. The ideal case is that the hive's entrance should face southwest so that the bees can use the sun to return to their hives.

You can use fences, hedges, or trees to protect the bees from solid winds. Another crucial practice is placing the hive around 18 inches above the ground level. This will protect the colony in case of floods or other land hazards.

The location and spacing are also critical factors to consider. These should be placed at least five feet apart to avoid overcrowding the bees and to allow enough space for them to fly around. Select a site that can be easily accessed for maintenance and monitoring of the bees, for example, a place close to the homestead.

Consider noise and quietness. It is also important to locate the hive in an area that is not too noisy to reduce stress on the bees. Noise can affect bees and slow down honey production. The best areas are not noisy, such as fields and rooftops located close to the roads.

Wind protection is another essential factor that needs to be considered. Ensure the hive is placed leaning onto a wall or a fence to protect it from the wind. Reflect on the direction of the wind and position the hive to reduce exposure to the wind.

Another thing to consider is safety from predators. To enhance the bees' ability to fly away from the hive in an upward direction to escape from the predators. The nets should be used to prevent birds from accessing them, while the holes should be small to prevent large animals such as bears or pets from getting to the hive.

Also, try to move your hive as little as possible. Do not transfer the hive often, as it will disorient the bees and make them stressed. If relocation is needed, it should be done when bees are least likely to be active, either at night or during the winter.

Providing Proper Ventilation

Controlling and ensuring the flow of fresh air is crucial for bees' healthy lives because it controls the hive's temperature and the amount of moisture that is likely to accumulate and cause diseases.

Ensure that the hive entrance size is correct for the colony. A small entrance hinders fresh air flow, while a large entrance exposes the hive to various invasions by petty predators. The entrance should be of a size that allows the bees to come and go without difficulty but one that intruders cannot easily access.

Some hives have tiny openings or meshes at the side or at the top and bottom to allow air to get in and out. These can be mounted on the two sides or at the top of the hive. They should not let pests into the house, but the holes should be large enough to let in fresh air.

Installing a screened bottom board is suggested because it provides better air circulation and assists in managing mites and other pests. The screen should have a mesh on it to ensure that no bees fall through because of the size of the mesh.

Check the hive frequently to ensure adequate airflow and adjust the ventilation if necessary. Look for signs of too much humidity or bad ventilation, such as droplets of water on the walls.

Protecting from Predators

Protecting your bees and their home is essential because other creatures can invade the hives and cause harm. When large predators such as bears or raccoons are around, it is advisable to put a fence or barrier around the hive.

The fence should also be sturdy, and ideally, it should be 3 feet high to discourage children from climbing on it. Use entrance reducers to ensure that it becomes hard for predators to penetrate into the hive. One should also place the hive on a platform so that it becomes hard for predators to access it.

Other actions that can be taken are fixing lights or alarms that go off whenever animals are active at night to scare the night predators. It is also vital that you occasionally inspect your hive and look for things such as claw marks or frames eaten by the predators, among others.

If you find out that the hive is unhealthy, you can do certain things to improve it. You can also add a frame of brood comb in the hive as you do it also. This is the comb that has eggs, larvae, and pupae. The brood comb will help the hive regain numbers or, in other words, replenish the hive.

Also, you can make them take sugar syrup, or you can offer them pollen patties. Sugar syrup, also known as feed, is a sugar and water solution which is the food given to the bees. Pollen patties consist of pollen and provide the bees with protein.

Part 4: Colony Care and Management

Chapter 6:
Installing Your Bee Colony

The construction work is done. Now it's time to get to the real deal: introducing and managing bees. This chapter is all about that. But you also have to make sure that the bees don't hurt you. So, you'll need some protective gear for any kind of interaction with the bees.

The Protective Gear You Need

Protective clothing and equipment should be of high quality to avoid being stung by the bees and to control them effectively. Here's a comprehensive, detailed overview of the essential protective equipment for beekeeping:

Bee Suit

A bee suit is essential for complete protection, and it covers the beekeeper from the head down to the ankles. A good bee suit is made of thick cotton or any other similar material that is breathable and protects the wearer from bee stings while being comfortable to wear. Some of the features that are considered important are the zippers that should be strong, elastic bands around the wrists and ankles, and reinforcement of areas that are likely to be stressed, such as the knees and the elbows. Such suits are especially useful in hot weather because they do not overheat the wearer and allow the beekeeper to work long. Ventilated suits have small holes in them, making them the best for summer since they allow air circulation while protecting the wearer.

Veils

Veils are essential because the face and neck area are the most sensitive to bee stings. Usually, veils are made of mesh fabric that enables air circulation and visibility while keeping the bees away from your skin. They are categorized into round veils, fencing veils, and hooded veils. Round veils are worn around the head and create a cylinder, which gives a good field of vision and enough space to avoid touching the face with the veil. Fencing veils have a vertical front element, which is usually chosen for comfort. Hooded veils are part of the bee suit and do not have any gaps where bees can get in. It is important that the veil should be sufficiently firm to keep it out of the face to avoid stings through the mesh.

Gloves

Gloves are very important for protecting the hands. Gloves can be of different types, and each type has its own level of protection and flexibility of movement. Leather and goatskin gloves provide good protection and are very durable but limit the hand's flexibility. They are perfect for beekeepers who do not mind the lack of precision in handling the smoker as they value their safety. Goatskin gloves are especially preferred due to the fact that they are very flexible and comfortable to wear. Nitrile gloves are more sensitive and flexible than other types of gloves, and they enable beekeepers to work on frame inspections more efficiently. But they are not as protective as leather gloves when it comes to stings. Air-permeable gloves that do not allow sweat and moisture to accumulate and make the hands hot during warm weather. It is important to inspect and replace the gloves frequently in order to ensure that they are still effective.

Smoker

A smoker is an essential piece of equipment used when inspecting bees as it helps calm them down. The smoke masks the alarm pheromones that the bees release, and this makes them less aggressive and hence can be easily dealt with. A smoker should be used in a proper way, for example, using pine needles, burlap, or cardboard to create soft white smoke, which is harmless for bees and the beekeeper. The smoker should have good bellows and a heat shield to avoid getting burnt. It is also advantageous to have long-burning fuel that will give out a constant column of smoke and is easy to clean.

Hive Tool

The hive tool is used in many ways, such as separating the hive components, scraping off the propolis, and lifting the frames. A good hive tool should be made of high-quality steel so that it can be used several times without bending or breaking. It should have different edges and prying points to enable it to perform different tasks within the hive; thus, it should be a multipurpose tool for routine hive inspections and maintenance. Comfortable handling is also facilitated by ergonomic design to minimize hand fatigue when the equipment is used for a long time.

Additional Protective Gear

Apart from the basic equipment, there are other items that can be useful in increasing safety and productivity. Boots that are of good quality and comfortable to wear and extend above the ankle help to keep the bees out and offer additional cover. They should be strong and appropriate for the area where the hives are placed. Beekeeping aprons can help save your clothes and also add extra pockets for the tools, which can be convenient during the inspections. It is also advisable to wear a light hooded jacket for additional covering, which helps soothe the bees and avoid stings.

Protective Clothing Maintenance

It is important to keep your protective gear in good condition so that it can continue to protect you and be effective. Check your gear often for any signs of wear and tear, including holes, loose stitching, and damaged zippers. Fix any damage on the spot with the right material, such as patching tape or any spare part that may be required. If your gear is worn out, it is advisable to replace it to prevent being stung by bees in the process of shooting.

Hence, beekeepers should ensure that they purchase good quality protective clothing and equipment and ensure that they are well maintained to minimize the dangers of handling bee colonies. The right equipment also safeguards you and assists in managing your bees in the best way possible for the safety of both you and the bees. Now that you know all about the necessary protective gear, let's learn how to introduce bees into the hives.

Safely Introducing the Bees to the Hive

You have your bees, all the equipment you need, and are all suited up in your protective clothing. It is time to put your bees in. Yes, please put on your protective gear. The whole process can take about two hours, so it is advisable to take your time and be slow and careful throughout.

If you have everything in a good position, then you can begin. Firstly, feed your bees with sugar water. For this, a new bottle is advised, and the solution is made by combining equal parts of water and sugar. A shower of sugar water makes the bees feed, and bees with full stomachs are less aggressive. Also, a package of bees typically does not contain a hive, honey, or a queen to guard; hence, most are usually tamed.

Next, remove all frames in your hive body and dip the two sides of the foundation into the sugar solution. Wash the frames, after which you should replace them on the circuits; you only need to spray each side a few times. This is the first time you will physically interact with your bees, so you should not get nervous or panicky and be very slow. Like any wild animal, bees are sensitive to abrupt movements, so one should stroll when they are around.

If it has been a while since you sprayed the bees, you should do that again for better results. The next step involves using a hive tool to gently remove the sugar water container without spilling the queen cage that is usually placed inside the package. After pulling the canister out, put it on the frames above, shaking off the sticking bees into the hive.

Delicately take out the queen's cage from the packaging. In this case, different companies employ various techniques to release the queen. Some companies offer a capped tube with the sugar paste. Twist off the top, take a portion of the paste, and put it in a form that the queen can eat through, and after several days, she will be released into the hive. This means that the queen must stay in the cage within the hive for several days to release the pheromones that are used to show that she is the queen.

Take a wire, place it around the cage hole, and fasten it to any of the frames present in the hive. After three days, it is possible to release her and unhook the cage to check if she has been released. If the company does not provide paste, you can put a miniature marshmallow in the tube after removing the cap; the queen and bees will have to chew through it within a few days.

Most bees will be clustered within the frames in roughly an hour. Here, you can place one pollen cake over the frames and put on the hive cover and the telescoping cover over the hive. If you leave the hive where you installed the package, return the bees to their hive and feed them with sugar syrup using the entrance feeder before securing it with tape.

Lastly, add the entrance reducer, and it is ready. If the hive is too close to your home or needs to be relocated, use a piece of tape to close the entrance and cover and then transfer the entire hive to the new location. After arriving, remove the tape on the entrance. Placing bricks on the top of each hive can help prevent raccoon and skunk invasions.

It is recommended that you release your queen after approximately three days of installing your bees. If possible, do this without lifting the hive cover so as not to disturb the hive.

Give them a treat of sugar water syrup, which should be there for the initial month because bees will drink it a lot. Ten days later, a standard hive inspection was performed to determine whether the queen was laying eggs and whether the bees had started constructing a comb.

Initial Colony Care

After establishing a colony, feeding, and varroa mites are the two major things to concentrate on. Educating yourself on how to care for them is essential because a honeybee colony is alive. Organized beekeeping has become complicated since the arrival of the parasitic varroa mite in early 1990.

Apart from varroa, other stress factors include persistent European foulbrood, loss of forage, climate change, low nutritional quality of pollen due to rising CO_2, and possible pesticide exposure in regions with intensive commercial beekeeping.

To ensure the health of your beehive, it is crucial to do regular inspections of the hives. Hive inspection on a sunny day should not involve much stinging. Master the use of smoke in a moderate and limited manner to influence the bees' actions without disturbing them. Be sure to wear a veil to shield yourself from the painful stings, especially on the head part of your body.

You should also bring a lightweight hooded jacket for further covering. Used in calming bees, the smoker should give out white, calm, and dense smoke. A hive tool is used for wedging frames apart and scraping off propolis. It is recommended that the offset be half an inch. You can use vinyl or goatskin gloves while inspecting.

The bees should not be provoked or threatened in any way, so one should move slowly towards the hive. Smoke occasionally to subdue guard bees at the entrance and top bars of the hive. Never attempt to take a frame if bees are observing you. Only move forward when they are not looking in your direction.

If bees start bumping or stinging, one should stop and wait until the bees are out of the angry mood. Look for varroa mites in the colony from late June to later. Avoid being too rough when smoking, and ensure that you do not overhandle the food. Observe the bees' reactions and adapt the actions accordingly to the bees' reactions. If the bees get too defensive, seal the hive and attempt to take samples at a different time.

It is tough to comprehend the social behavior of bees, which depends on many factors, such as parasite invasion, climate change, and habitat change. The bees need the beekeepers' services as the colony's caretakers and protectors to ensure that they are balanced. Beekeeping entails routine checking of the hives for diseases or pests and other routine management practices. Having a clear strategy and plan is essential when setting up hives to achieve the best results.

To maintain healthy bee colonies, beekeepers have to remain alert and actively prevent diseases that threaten bee populations. Several factors must be considered in order to obtain honey successfully, including determining the right time for honey collection, the proper way of frame removal, honey extraction, and storage, and practicing proper and sustainable methods of honey collection.

Following these guidelines and ensuring that you keep learning will put you in a well-placed position to manage your initial bee colony correctly. Management of the hives, wearing correct protective clothing while handling bees, and controlling varroa mites are some critical factors of beekeeping.

Common Challenges and How to Overcome Them

Beekeeping can be satisfying and profitable, though some difficulties may appear. Here are some common problems faced in beekeeping:

A significant challenge is the presence of such threats as pests and diseases that can harm honeybee colonies. Some of the many pests and diseases that can severely affect bee colonies include Varroa mites, small hive beetles, wax moths, and foulbrood disease. Integrated Pest Management (IPM) is crucial for controlling these issues. This includes regular hive inspections and the use of specific treatments such as formic acid, thymol, or oxalic acid, which are effective against Varroa mites under certain temperature conditions. For wax moths, natural predators like chickens and certain wasps can be made to roam near hives to help control moth populations.

Another factor that can be problematic is weather changes and other instances of extreme climate. Temperature plays a crucial role in the life of bees and any changes in the weather. It can affect their ability to go out in search of food and to protect the hive. To mitigate these effects, beekeepers can insulate hives against extreme temperatures and ensure hives are shielded from direct elements like direct sunlight.

Lack of adequate food in the surrounding environment of the beehives will cause malnutrition among the bees. These deficits can result from urbanization, pesticide application, and monoculture agriculture, which limit access to various and plentiful nectar and pollen sources. Planting a variety of pollen-rich plants and providing supplemental nutrition during scarce periods can support bee health. Also, be mindful of the impact of pesticides on bees' health.

There are certain conditions characteristic of different seasons. For example, we control swarming in spring and provide bees with food and moisture in winter. This is why it is essential to be aware of the risks associated with nearby agricultural use of pesticides and the proper location of hives. Understanding these seasonal patterns and preparing in advance can help beekeepers maintain healthy colonies throughout the year. This might include adjusting the placement of hives based on seasonal pesticide usage in nearby agriculture, which can drastically reduce the risks of chemical exposure to the bees. The next chapter will discuss these seasonal changes and management in detail. So don't worry, we'll cover every aspect of it.

The queen bee is essential in the honeybee colony, and any problem that might affect the queen will affect the colony. Abnormalities like an unwell queen, queen supersedure, or queen loss jeopardize the colony's reproductive system, leading to low honey production and even colony collapse. Regularly monitoring the queen's health and replacing her when necessary is important to ensure the stability of the colony.

Colony Collapse Disorder (CCD) is a disease that affects honey bees. In this disease, workers in a bee colony exit the hive and die. This needs to be worked on constantly, and management strategies are required. Managing CCD requires a comprehensive approach that includes maintaining strong genetic diversity, minimizing stressors such as pesticides, and ensuring optimal nutrition and hive management practices.

Beekeeping is an exciting and, at the same time, a rather tricky occupation. It is crucial to recognize these issues and work to manage them to maintain healthy and productive colonies of honeybees. The beekeeper must also keep updating himself/herself on new developments in best practices and emerging challenges in the apiculture industry through information-sharing forums and training.

Chapter 7:
Seasonal Hive Management

It is important to understand that the methods of managing a beehive change depending on the season or period in question. Every season has its own management techniques that need to be implemented to support the colonies' survival and productivity.

Spring: Preparing for a New Season

Some management practices should be followed for honey bees during the spring season. This is the time when cleaning is necessary inside the beehive, as bees can be starving with no food and flowers have not yet bloomed. They tend to swarm if they do not find what they seek, so timing is crucial.

This is the first step to ensure that you are well-prepared to face the challenges associated with late winter and early spring. Inspect all your beekeeping equipment, especially those stored away for some time. Replace or repair whenever necessary; the best thing is to order some of the supplies before the busy season begins.

Observe the entrance of the hive to determine its condition and needs. In hot weather, look inside for a minute. Temperatures can differ slightly, so ensure it is warm enough to conduct a short inspection. The queen's presence and eggs that have not been laid for long should also be sought. There should be a good worker brood pattern under good rearing conditions and with a healthy queen. If there is no brood or only drone brood, it is best to buy a new queen.

If the weather is favorable, these bees will forage for nectar and pollen. It is advisable to monitor the food stores frequently because a colony may survive the winter only to die from hunger in the early spring. If necessary, feed the bees with sugar water or provide protein patties. Mix pollen and make patties, and then put them in the hive.

As spring continues, the size of the brood nest increases. If bees have swarmed and landed on the topmost boxes, the honey is stored in the top box while the brood nest is relocated to the lower box. A new box should be added if the bees need more space to store honey.

Swarming is a natural process but not always palatable to beekeepers. To avoid swarming, one can split colonies or create small nuc colonies. These can be used as small hives and can be expanded into full-sized hives or combined with others at a later time.

Bacterial or fungal infections, such as nosema or chalkbrood, may be evident in spring, but improved foraging conditions can alleviate them. Perform Varroa mite inspections and mite drop counts frequently. Some treatments require specific temperatures that should be noted or incompatible with the honey collection, so adjust accordingly. Document the hive conditions and tasks to monitor progress and prevent repeating mistakes.

This is evident in the later part of spring when flowers are plentiful, and young bees are produced. Watch for queen cells to prevent swarming; this weakens the colony and causes conflict with neighbors. Swarm control can be achieved through the use of frames, supers, and even by eliminating queen cells. Division of colonies can also be used to control swarming instincts and elevate colony numbers.

Summer: Managing Population and Health

In summer, flowers are in full bloom for honey collection, and winter stocks are needed. New queens come of age and start laying eggs, and honey is typically collected. Be alert for Varroa mites, particularly during the peak of brood rearing. Mites result in sicknesses and disease transference, which is hazardous to the colony. Treat mite infestations promptly.

Nectar becomes scarce in late summer, and this makes bees aggressive and indulge in robbing. Reduce the number of entrances and exits in the hives and mount robbing strips. Winter bees are reared during autumn, and they live longer and have larger fat bodies to support them during winter. Mite Management can be done through Integrated Pest Management (IPM), which assists in regulating the density of mites while not affecting bees or the surrounding ecology.

Some of the practices that farmers do during summer include opening the hives to assess functionality, forage, nectar and pollen, capped honey, and brood. Look for diseases such as Varroa mites and wax moth larvae. Proper measures should be employed to deal with wax moths and avoid the destruction of hives, including the use of screens. There should be a provision of lay supers during concentrated nectar flows to prevent crowding and swarming. The amount of honey in the hive should be checked to determine when to harvest.

Fall: Harvesting and Preparing for Winter

Hive management helps check the health of colonies and prepare them for the winter period. Two inspections, the first in early September and the second in mid-October, confirm the colonies' disease-free status, food supply, and available space for winter feeding.

The position of the colony inside the hive and its dimensions are the key factors determining how the bees will winter. A cluster of bees should be located in the middle of the bottom brood box, surrounded by frames of pollen and honeycombs. If bees are mostly in the top box, swap the boxes' positions or decrease the brood nest's area. If required, portions of the feed should be given out to the smaller groups. Syruping fondant feeding or sugar feeding is usually practiced in the fall.

Swop frames between solid and weak hives to balance them out. Do not cross-infect from hive to hive. Pair weak and stronger colonies to make up the experimental groups if necessary. It is essential to check on honey stores and mite loads to avoid losses during winter.

Prepare for winter by making your entrances small enough to deter robbing and pests. Provide feed to the bees during inspection since they need to survive. The second inspection involves checking honey stores, resizing hives, and equalizing or balancing the colonies. Use adequate insulation and ventilation to address the issues of cold and moisture that may develop.

Winter: Ensuring Colony Survival

Winter management involves ensuring that the hives are well insulated to provide the right temperatures for the bees and provide them with feed. Weigh the hives to determine their weight and check if they have enough honey supplies. Provide the bees with sugar syrup, fondant, or any other food if they require it.

Bees huddle around the queen for her to warm them. Ventilate the area properly to avoid the formation of moisture on the walls. Attend beekeeper meetings to be updated on which practices should be adopted. This way, the colony is protected and maintained until spring to continue the reproduction cycle.

These practices are essential for the beekeeper to ensure the colonies remain productive throughout the year in response to seasonal changes. As in any other farming activity, bee farmers must adapt to this natural occurrence by adequately preparing their bees during every season and closely monitoring them.

Chapter 8: Hive Inspection and Maintenance

Without proper inspection and management, your colonies will die. As we mentioned before, beekeeping is the hobby of a responsible person. We have touched on this topic in previous chapters; we'll go more deeply this time.

Regular Inspection Techniques

It is very important to inspect your beehive occasionally to keep your colony healthy and productive. Such checks allow you to visually observe the state of the hive, identify issues that may arise or already exist, and, if necessary, correct them to ensure the stable functioning of the bees. Below is a step-by-step guide to performing a proper hive inspection.

Before starting the inspection, ensure you have all necessary equipment: Hive tool, smoker, protective wear (jacket, gloves, veil), and bee brush.

The initial process of inspection entails smoking the hive. Calm the bees by using the smoker to blow some smoke at the entrance and under the lid. This conceals the alarm pheromones and calms down the bees, making them less aggressive. Second, lift the outer cover gently and then the inner cover to reveal the entire hive without disturbing the bees too much.

The brood must be inspected during the process because the condition of the brood is a sign of the colony's condition. The basic pattern of the brood is also essential. You need to have eggs, larvae, and capped broods. Healthy brood patterns will, therefore, have most of the cells occupied, meaning that the queen is productive and healthy.

This shows that there is a healthy brood, which can be pointed out by characteristics like uniformity of the brood pattern, visibility of eggs, larvae, capped brood, and very few drone cells at the outer fringes of the brood. Disruptions are manifested in a dispersed pattern of the brood, a large number of empty combs, a large number of drone cells, and the existence of queen cups or queen cells.

The other important task is checking honey and pollen stocks. They check for sufficient stocks of honey and pollen to support the nutritional needs of the bees and to assist in brood rearing. They observed that frames are filled with capped honey and cells full of pollen in healthy stores. Emergencies are marked by low honey deposits and poor pollen supplies.

Pests and diseases are also vital areas that should be inspected. Varroa mites are one of the most dangerous pests affecting bees, so it is crucial to look for them regularly and treat the bees accordingly. Search for mites on bee bodies and count using a sticky board or the sugar shake technique. Look for signs or wax moths since these pests will likely compromise the comb and weaken the hive.

Supering is placing additional boxes (supers) to ensure the colony has enough space to store honey, especially during the honey flow seasons. This aids in discouraging swarming and encourages proper honey production. Super when nectar flow looks good, and the supers should be weighed to check when they are full.

When inspecting the equipment, look at the comb and its disrepair, if any. A light-colored comb is considered healthy, while a dark and dirty comb may be a sign of age and should be replaced. Check the burr comb (irregular comb) and remove it if it interferes with the functioning of the organ. Switching to black or dark honeycomb for the next season would also be advisable.

The best time to check a hive is in the morning when bees are in the field searching for food. Temperature variation, with 60°F (15.6°C) to 90°F (32.2°C) being preferred during the midday, is also recommended. However, it is recommended to refrain from inspecting during a rainy day or when the temperatures are too high or too low.

New colonies should be inspected every two weeks, especially during spring and summer. For well-established hives, a monthly check is enough, as the bees are capable of tending to the hive on their own. Please do not open the hive during winter unless you have to when checking the entrance area to ensure it is not blocked by snow.

Some important points to note during an inspection include avoiding noise and swift movements to ensure that bees do not get agitated, approaching the hives from the rear or the side, using smoke only when necessary and sparingly, cleaning wax debris to prevent the attraction of pests, and using a pencil to mark frames of interest for later use.

Besides, it is crucial to pay attention to the bees' activity outside and inside the hive. Monitor for dead bees being taken out of the hive by the bees themselves, which indicates cleanliness. Look for robbing behavior, especially during nectar shortages, and look at the activity at the hive's entrance to determine the colony's health.

Regular inspection of hives allows for collecting essential data regarding the health and output of the bees. You can keep your apiary healthy and productive through these points of practice, accompanied by necessary preparation and appropriate technique. Have fun, and be thrilled with beekeeping as a fulfilling practice!

Identifying and Addressing Issues

Examining a beehive to have a prosperous and healthy functioning bee population is essential. A successful inspection requires comparing the hive's health signs with signs that may need attention. This guide lists the main things that should be checked during the inspection of a beehive and includes recommendations on dealing with some of the issues.

New beekeepers typically want to check on the hive frequently after starting with a new hive. Early inspections should be done within one week, and the frequency should be gradually reduced to avoid stressing the bees.

Sanitation is essential to the bees and their habitat since it will help them maintain their hives. Some hygiene aspects of monitoring are Cross Comb and Burr Comb.

These irregular comb structures can pose several challenges when they are poorly managed. Properly cleaning them makes hive management easier and minimizes the likelihood of harming the bees during frame extractions.

Cross comb and burr comb must be removed gently so that one cannot develop in the future again. This way, cleaning these structures is more manageable and healthier, allowing for further efficient inspections. Do not pour sugar syrup carelessly around the hive, as it may attract pests and robber bees.

You should always ensure that you leave that hive in a better state than you found it. A beekeeper may leave tools inside the hive or spill sugar syrup, which may attract pests and robbers.

Other Issues to look out for are a left or right skewed brood pattern, too many drone cells, and the appearance of queen cells suggesting swarming or queen rearing.

Pests and diseases pose a great threat to bee colonies and, therefore, should be monitored to avoid the unfortunate event of colony collapse. Monitor for varroa mites on a regular basis. These conditions, if diagnosed early enough and treated, will make a big difference. Check for signs of webbing and tunnels in the comb, as these are signs of infestation by wax moths.

Another beekeeper's tool is a smoker, which is used to calm the bees when conducting an inspection of the hive. It is important to blunt the alarm pheromones and make the bees calm, and to do this, a few gentle puffs at the entrance and under the lid are used.

Check hives when the weather is warm, and bees are foraging or, preferably, when not flying. Do not inspect the bees during rainy or extremely hot and cold weather since this will cause stress to the bees, who will be busy.

Observation from outside can help without interfering with its activities, which is precisely what the beekeeper did. Check for signs of dead bees since bees will remove a dead member, which is an indicator of cleanliness. A high-activity rush by bees at the hive's entrance or a different colony is a sign of robbing.

On hot days, bees use their wings to help in cooling the hive with the help of their wings. Ventilation should be considered in this process and avoided to prevent the car from overheating.

Strictly ensure that all feeding materials are clean and placed inside the hive to avoid attracting more visitors. Create a checklist to help prevent tools from being left behind after an inspection. This small act can help avoid developing other issues in the future and keep the hives in order.

Observing bees and opening the hives is an interesting experience that allows us to evaluate the condition of the colonies. Bee-rearing is one of the most satisfying practices that one can undertake, and routine checks are one way of having a productive apiary.

Part 5:
Sweet Harvests

Chapter 9: Honey Harvesting

When to Harvest Honey

Generally, beekeepers collect honey at the end of a significant nectar flow and when the beehive contains cured and capped honey. The conditions and circumstances differ from one region to another. New beekeepers are fortunate if they can harvest a small amount of honey by the end of the summer. This is because a new colony takes a whole season to develop a large population to enable it to harvest a surplus of honey.

Inspect the hive every two weeks during the summer to check on the progress of the honey bees. See your bees' progress and determine how many frames are filled with capped honey. If a shallow frame has 80 percent or more of sealed, capped honey, please feel free to take out and take this frame. Or, you can wait until one of the following is true:

The bees have occupied all the frames with capped honey, and the final major nectar flow of the year has been finished. If the honey in the open cells, which are not capped with wax, is cured, it can be harvested. To check whether it has cured, the frame with the cells should be placed upside down. Try tapping the frame lightly with your hands. If honey is found to be leaking from the cells, it is not cured and should not be extracted.

This is not even honey; it is a mixture of sugar and other chemicals. It is the nectar that is not fermented. The moisture content is too high for it to be classified as honey. If the nectar is preserved in a bottle, the outcome is a watery syrup that can quickly ferment and go bad.

You must wait until the bees have collected all the honey they possibly can, so be patient. That's a virtue. But do not let the honey supers, i.e., the boxes containing the frame, remain in the hive for too long! Apart from dedicating a weekend to harvesting your honey, you may have many other things to attend to. But do not delay what has to be done. If you wait too long, something undesirable can occur.

When the final primary nectar flow occurs and winter is on the way, bees start consuming the honey they have produced. If you let supers sit in the hive long enough, the bees will consume most of the honey you intended to take. Or they will transfer it to the open spaces in the lower deep hive bodies. Either way, you have lost the honey that should have been yours.

Do not let it get to that point; remove those supers from the hive! If you delay pulling your supers, the weather becomes too cold to harvest your honey. During winter, honey becomes thick or may crystallize, a factor that makes it difficult to remove from the comb. Remember that honey is most accessible to harvest when it is still warm from the summer and can be easily extracted.

Extracting and Processing Honey

When preparing to extract honey, prepare the necessary tools, as things may become messy once the process begins. Some of the tools that you will require include a table, honey extractor, 400–600-micron filter, plastic hand gloves, a bucket for honey, an extra tub for wax cappings, clean jars, a space heater, a heated electric knife, a capping scratcher (optional) and cheesecloth.

First, there is the extraction process, where wax cappings are removed from the honeycomb. Beekeepers prefer to use nine frames instead of ten in the supers so that the bees have enough space to draw the comb out, and hence, it becomes easier to pull off the wax caps. Wax cappings should be removed from the honey frames using a heated electric knife.

Place it above the honey bucket to catch the wax as it melts off the frame. Cut from the top of the frame to the bottom using the hot knife while being gentle to avoid scrapping the frames. If using a metal knife, it is advisable to occasionally warm the blade by immersing it in hot water.

Some beekeepers use a bread knife, and others use an electric knife. To extract honey from wax caps, drain them through cheesecloth into a bucket. The cheesecloth filters out the bees' wax caps, leaving you with the honey you want. Beeswax caps can also be sold to make candles and other products.

To get liquid honey, use a centrifugal extractor. Balance the frames in the extractor so that the weight does not shift, causing unnecessary vibrations. Place a honey bucket and filter close to the spout so that it can catch the honey. When the extractor rotates, honey will be poured out through the spout.

After one side is complete, shut off the machine, turn the frames, and continue until all honey is extracted. For those who do not have an extractor, let the honey drip through the uncapped frames into a bucket in a room that is at least 80 degrees at night. However, using an extractor is the best way, and you can usually rent one from a local bee club.

Do not take honey from a beehive that is being treated with antibiotics or any other kind of medication. If you have to treat the hive, do it when there is no honey in the combs or when it is still young, weeks before or after the honey flow.

Storing and Packaging Your Honey

Honey can be stored in different forms, including liquid and granulated forms. It may be stored in a honeycomb until the bees are prepared to use it, or they may remove the honey and store it in different receptacles. The honeycomb structure used in storing honey differs slightly from that used in storing liquid honey.

Preparation for Storing Honey

Preparations are necessary when it comes to storing honey. You should also make space to store your honey because honey is an essential product. It could be a cooled space or just a room in your house that is not exposed to large temperature fluctuations.

Temperature and Light Considerations

Cool temperatures are known to cause honey to crystallize. This is normal and can be reversed. Consumption of crystallized honey is not a problem. It can be returned to its more liquid state by heating it gently.

Excessive heating of crystallized honey hurts its nutrient content. You should also not expose honey to more than one crystallization and warming cycle. If you do, you will be left with honey with its sweetening ability but no micro-nutrients and volatile compounds that are in it.

Honey should be stored in a cool, dry place away from direct sunlight. The temperature at which it is stored determines its crystallization, which may or may not occur. When storing honey, ensure that it is not in close proximity to appliances that generate heat, such as stoves. If your house becomes warm during the day, try to locate the coolest area and use it for honey storage. Honey does not need to be stored in a refrigerator and has a very long shelf life. It is also easier to handle honey stored at room temperature than that stored at a cold temperature.

Storage Containers

Food-safe plastic and glass are the best and most preferred storage container materials for honey. Earlier, honey was stored in earthenware pots that had a sealable lid or something similar. It is advisable to store honey in a plastic container with a lid or a glass container with a lid. The containers where you will store honey should also be clean and dry before you place honey in them. Homemakers and beekeepers prefer using glass mason jars to store their honey. These jars enable one to view the appearance of the honey in the jar as well as the process of crystallization.

Much concern has been raised about the use of metal containers in storing foods. Even though food-grade metals can come into contact with food for human consumption, using metal containers to store honey is discouraged. Honey, when in contact with metal surfaces for long periods, undergoes oxidation, imparting a rather nasty metallic smell and taste. Determining whether such honey is safe for consumption is not very straightforward. On the same note, metals are also used in honey collection. The knives used to cut off the caps of the honeycombs are made of metal, as are the honey collectors, which consist of metallic drums. However, the time that these metal parts or equipment are in contact with honey is limited and not sufficient for the oxidation process to occur.

Freezing Honey

Freezing is also employed in honey storage. To freeze honey, transfer it to a container and ensure enough space for the honey to expand. Put the container containing honey in the freezer, and the honey will remain fresh for a long time. Frozen honey is not solid because it contains deficient moisture. Honey is mainly sugars in solution with 18% moisture or less. This does not allow honey to freeze to a solid mass but is very sticky and thick. When defrosting honey, it is recommended that one does not apply heat directly on the honey. Honey should be thawed using heat in a roundabout way because honey tends to crystallize. Submerge the container with honey in warm water and let it soak for a while. The heat from the water softens the honey, making it less thick. If the water cools down too much with the cold frozen honey you are adding to it, you may heat it. Do not thaw the honey in large quantities; only thaw the portion of honey that you are going to use within a short period of time.

Light Exposure

There's an opinion that honey stored in containers and exposed to light is not the same as honey that has not been exposed to light. It is used to make the honey turn brown. Also, exposure to light and heat, such as sunlight, may cause the honey to heat up and lead to higher levels of Hydroxymethylfurfural (HMF), which isn't a good thing to consume. If you are not going to freeze honey, the most appropriate place to store it is in a cupboard or kitchen pantry. You should never put some things in the refrigerator; honey is one of them. This is so because refrigerators undergo large temperature fluctuations, which are unsuitable for stored honey.

Chapter 10: Working with Beeswax

This chapter will focus on what remains in the hives once honey has been taken from it or extracted from it. The beeswax is also a useful product that is utilized in the production of many various items.

Harvesting Beeswax

Extracting beeswax is best done when the bees are least present in the hive, especially in the evening when they are out looking for nectar. This reduces the likelihood of getting stung and assists one in conveniently accessing the frames in the nest.

Dress up like a beekeeper in your protective gear so you do not get stung by bees while taking pictures. Ensure that your face is enclosed within the mesh and that the suit properly fits at the back. If you do not have a beekeeper's outfit, then dress in thick cloth covering your body and wear a hat with a veil to cover your face. Do not let the mesh contact your face since this leads to stinging.

Calm the bees using a smoker so that you can easily access the hive. The smoke from the smoker causes the bees to descend to the bottom of the hive and leave the honeycomb, making it easy to take the frames. For this purpose, you need to use a bee smoker. Using a bee smoker effectively and safely is crucial to minimize stress on your bees. Select smoker fuels like pine needles, dried grass, burlap, or untreated wood chips for their gentle, white smoke. Start by lighting kindling, such as crumpled paper or dry leaves at the bottom, then gradually add your chosen fuel, keeping it loosely layered for air circulation.

Use the bellows to maintain steady smoke. Test the smoke's temperature by puffing some onto your skin; it should feel cool. If it's too hot, adjust your materials or add more fuel. Gently puff smoke at the hive entrance and under the hive cover to mask pheromones and reduce defensive behavior. Use smoke sparingly during hive inspections, targeting areas where bees cluster to keep them calm. After use, extinguish the smoker properly by closing it off and allowing the materials to burn out completely in a safe area.

After smoking the bees and removing the frames, use a hot, uncapping knife to cut and peel off the wax cover from the honeycomb. With two hands, grab the frame, and using the knife, run it along the frame from the top to the bottom. The wax cappings should be relatively easy to remove if one applies moderate pressure to them.

Once you have stripped the wax cappings, put the caps in a bucket and let them stand for 15-30 minutes so that the honey can drip from the wax caps on its own. Then, remove the wax caps and put them in a separate bowl for melting in a process called rendering.

To prepare the wax caps, they should be dipped in a thin layer of cheesecloth and placed in a large pot of boiling water. Melt the wax and beeswax in a pot containing water on medium heat until the beeswax leaves the wax caps. Twirl tongs to wring the cheesecloth bundle and apply pressure to extract more beeswax from the mix.

Once all the beeswax has been melted, drain the cheesecloth and let the beeswax cool and consolidate. Sieve and drain off the remaining water and allow the beeswax to cool in a polythene bag or any other suitable vessel.

Uses of Beeswax

Beeswax has many uses in nature, and although edible, it is more common in washing and grooming-related activities. However, some of the ingredients of beeswax have moisturizing properties, which may help firm up the skin and make it look puffy.

It is noncomedogenic and anti-inflammatory, making it suitable for sensitive skin types, such as rosacea or eczema. When applied as a topical treatment, beeswax forms a barrier on the skin that shields it from harsh weather conditions and irritants.

There are many uses of beeswax that have been documented in the past, as illustrated in this paper. The first time beeswax was used was when it was shown to have been used to fill a cavity of a tooth that dates back to over 6,500 years from Europe and was discovered by archaeologists. It was also used in other prehistoric civilizations; for example, it was used in Chinese nail polish and medicine in 2500 BC and in hair treatment, painting, and mummification in Egypt.

Beeswax has also been discovered on clay vessels dating back to the 5th millennium BC, where it was probably employed as a varnish that made the clay pots waterproof. Greeks and Romans used beeswax with olive oil and herbs and incorporated it into salves for treating sores, cuts, and burns or into cosmetic creams for hydrating the skin.

Since the beginning of writing, beeswax has been applied to seal documents and letters. The use of beeswax in the lost wax method has been prevalent ever since metal casting came into the picture for making statues of bronze or jewelry of gold and silver. Beeswax candles have been discovered in Egyptian tombs and were used in the classical period in the majority of the Mediterranean region; the wealthy used candles, while the poor used oil lamps, which were cheaper. Candles made from beeswax were used in the Christian churches of the Middle Ages because they burned well and produced very little smoke.

Today, it is used in many different ways. It can be incorporated into solutions for making medical, cosmetic, and home care products, burned for making candles, consumed, liquefied, shaped into ornaments, and employed as a sealant.

Modern applications for beeswax include greasing door hinges, seasoning cast iron cookware, shining shoes, decorating Christmas trees, lighting dinner tables with beeswax smokeless candles, polishing furniture, conditioning cutting boards, making fire starters, blending it with pigments to produce encaustic painting, greasing of screws and in the making of lip balms among other uses. Beeswax remains a flexible material that remains relevant and useful in today's world and society.

Part 6: Transforming Hobby into Hustle

Chapter 11: Selling Your Honey and Beeswax

By now, you know everything you need to know to get started and sustain your beekeeping hobby. This next part is for those wanting to go big and make money from their hobby. This chapter will share various ways to monetize your beekeeping practices.

While reading this section, you might find the strategies mentioned here overwhelming, especially if you have no such experience. Don't worry; remember, you don't need to do it all simultaneously. Just start with something small that you're already comfortable with.

Let's start by discussing various marketing strategies you can use to reach the customer who'll buy your honey.

Marketing Strategies

To start with, it is crucial to concentrate on the process of searching for a specific market for your honey. Is it locally sourced? Is it perhaps because it is raw or organic? This will also help you understand the strategies that can be employed in marketing to your target market and come up with a brand message that will capture the attention of the intended group.

Modern customers are much more informed and care much more about the food they eat and its background than ever before. Remember to establish and be absolutely clear about your product's unique selling proposition and ensure that this is communicated persuasively and consistently across the marketing mix.

Another factor that is core to business is the community. Directly selling the honey products to the stores within your region or with the end users is advantageous to the honey business. It is recommended to cooperate with cafes, bakers, or even local markets where they can sell the product.

Direct methods of reaching out to potential customers are tasting events or organizing beekeeping lessons and explaining the role of bees in the environment. Apart from creating awareness of the brand, this strategy also makes you recognized as an expert in the industry.

The ability to market one's business, especially in the modern world where computers and the Internet play a significant role, cannot be overemphasized. Social networks are perfect for presenting products, especially if you are into illustration and bright visuals.

Instagram is perfect for posting bright honey photos or, in general, for giving the audience a glimpse of the honey-making process. Polls, stories, and live sessions can also be used to share interesting materials and create a base of active users connected to the brand on the Internet. Thus, the frequency and message should be consistent to maintain the audience and attract more people.

It is always fun to educate people while making them roll on the floor laughing. In this case, well-written articles, such as blog articles on the health benefits of honey or fun and educative videos on bee farming, can significantly increase web visibility and traffic to your site.

It makes the business or brand more believable and helps establish trust with possible consumers. The idea here is to turn this site into a store where people can purchase honey and the go-to source of information on honey and beekeeping.

Another sweet spot is email marketing if we focus on marketing your product or service. It also establishes a method of communicating with your customers to alert them to new products or services or to pass on any helpful information on beekeeping or the environment.

Ideally, it is also useful to categorize your email list based on the customer's interests or past purchase history to be even more targeted with your marketing. Such a move increases click rates and makes customers eager to have a glance at whatever you have to offer them.

Another way of publicizing your website (if you intend to have one) that cannot be overlooked is SEO or Search Engine Optimization. This ensures that if a particular individual is searching for products such as yours, he or she will be led to your store. It also assists in increasing the website's ranking. When doing the referencing, ensure that the keywords are also included in the website's content.

If the audience is to be targeted in certain regions, local SEO strategies should not be ignored; this includes the Google My Business listing for the area and customer reviews. SEO is another critical factor for growth as more people get to know your site and visit it.

But now let us share the last category, which, in my opinion, is crucial—the visuals. We eat first with our eyes, don't we? There are many incredible tools, such as Desygner or Canva, that can help create really professional-looking marketing materials even if the user has no design skills.

These tools are ideal for social media as they have templates for posts, banners, and many more, so it is easier to maintain the consistency of the branding on social media. Designer helps to quickly change designs for various campaigns or promotions, while everything will always look neat and well-groomed. In this regard, such professionalism aids in passing positive information to consumers concerning the quality of the product offered.

Most people do not know that you can also sell bees. Bees and bee products are needed in agriculture, hence the increased production and demand. If you have been able to produce healthy, hardy colonies, clients will come to you to purchase queens, drones, and workers to improve their own apiaries.

Of course, the bee business is not just buying and selling these hardworking little bees with wings. The byproducts of bees also have their significance. Honey, beeswax, propolis, and bee pollen are all saleable commodities, each having a central target market. In addition, original honey provides you a competitive edge in terms of price because it costs three or even five times more than the regular honey you buy from supermarkets.

It is also true that no business would be complete without proper marketing strategies to support its operations. Everyone knows that the process of beekeeping is not shielded from that rule as well. The best beekeepers have strategies that ensure people talk about their business; the brand story and even their marketing strategy are called digital marketing.

Finally, ethical standards are one of the critical aspects of any viable business in apiculture. Bees' health and safety must be preserved to maintain the economy and increase revenues to meet demands.

Like any other business, beekeeping does not happen overnight and takes time and energy; even occasionally, it is a test of entrepreneurship. But if planned well, with a love for beekeeping and the willingness to alter something in the environment, one can make it in the beekeeping business.

Setting Up an Online Store

Since you have learned how to go into the beekeeping business, it is wise to develop your idea further to compete with other players in the market. Researching the market will be beneficial for you even if, in your mind, you are very confident that you are offering a flawless product or service. Market research is essential for one to determine the customers, competitors, and the prevailing business environment in the market.

Here are some ideas for brainstorming your business name: Short and unique names are employed since it is believed that they are easily memorable and can be easily related to the product. The population has a disposition toward what is easily said and written. This must, in one way or the other, relate to your product or service offerings in the market. Friends and family can be your best recommendation sources, as well as your co-workers or even through social networks.

The employment of such words as 'honey bees' or 'beekeeping' will improve the search engine ranking or SERP. A location-based name helps establish a solid reputation for your business locally and is perfect for SEO, but it is not ideal for expansion. To make sure that the name you have chosen is not taken, you can go to the US Patent and Trademark Office website to check if the name you want to trademark is available for use or not, as well as the availability of the domain names associated with the name you have in mind using the Domain Name Search. Hence, it is wise to center on ".com" or ".org" as including any of them boosts trust. Finally, select the name from the list that has gone through this screening and go for registration of the domain and creation of social media pages.

Your business name is among those factors that distinguish your business from the others and make it unique. However, the moment the business begins branding, it becomes difficult for the business to switch the name. Thus, it is essential to be very cautious when deciding which business entity to undertake when forming a business entity. It is a fundamental step that should not be overlooked as it precedes paying taxes, financing, getting an account, and establishing a business.

There are now hundreds of choices available regarding e-commerce platforms, so here are some factors that will help you decide. You can use website builders like Shopify, Wix, and Squarespace if you are just starting out. You can also purchase hosting from websites like Hostinger and then develop your website using a tool such as WordPress. However, it should be pointed out that the platform selection is predicated on the organization's needs.

If you set up an online shop, you may have to pay additional costs. These include paid templates, applications, and related add-ons like email marketing plans. The general price will depend on the specific attributes needed for your website. For instance, if Wix does not support a specific checkout tool, you may have to purchase a tool that does.

This is very important because the online store's appearance determines its success. Research shows that a visitor perceives whether they like a site within 50 milliseconds; therefore, the need to grab their attention with an elegant e-commerce template. There should be a transparent bar at the top – the main category pages should be formatted clearly so that the user can quickly locate the information they seek. The website search bars are helpful for shoppers to find what they are looking for. Footer links that provide relevant information may be the 'Contact' page, the 'About' page, social media icons, and links to customer service. Word from the clients on the product that they have used – they are also the social proof that you are a real business selling authentic products. An example is the subscription form used to create the email list, which users can use to be notified of new products and other company information.

When uploading your products, it is always wise to first check the plan you subscribe to. Although most website builders will let you sell as many products as you want, some will require you to switch to a new plan first. For instance, Wix offers the option of selling up to 50,000 products, but this is subject to the condition that you use the core plan or a higher plan. If you intend to sell a few items, you can opt for a simple plan; for example, Hostinger offers a simple e-commerce plan that enables one to sell 500 products at most. Ensure that the builder and chosen plan are appropriate for your store.

Credit card payment using Visa and MasterCard dominates the payment systems on the internet. This is when a financial organization like Visa, American Express, Mastercard, etc, is associated with your online business. Customers can type their card information into the payment gateway's checkout service on the seller's site or be redirected to the payment gateway site. Almost all web builders can connect with payment processing systems. For instance, Shopify has options that are often selected, such as Stripe and PayPal, among others.

The second old-fashioned remuneration collection technique is using digital wallets. It is one of the easiest to implement, and these wallets have features that will help secure them; thus, they should be incorporated when designing an online store. As with the payment gateways, most website builders allow you to accept digital wallet payments for your online business.

Security is one of the most essential criteria for secure transactions. When choosing a website builder, the platform must have an SSL certificate, two-factor authentication, substantial customer login sections, and anti-fraud tools. To learn more about safeguarding your and your customers' information, please go to the website security checklist page.

Last but not least, about the payment section, you have to consider how you will be able to transport your products. You must take two steps beforehand: Create your shipping 'from' location. This is the address you will use when shipping your products. Ensure it is current so that all the shipping rates and taxes are correct to avoid changing them later. Select the type of shipment that you prefer. This is where you need to decide where you are shipping to, which means the countries, regions, or zones as they are known. This implies that each place you need to deliver to will alter the shipping price of each delivery, so the best approach should be to try and estimate each place's cost.

Some of the companies you must choose from for delivery include the following: The most preferred carriers are USPS, FedEx, and DHL Express. Most web hosting services offer these services. Shopify Shipping is a very convenient service that is provided in partnership with USPS, UPS, DHL Express, and FedEx. You can also obtain the coupon code for the shipping label and check the present rates.

It is also essential that the buyers are given as many choices as possible and that the options are not limited. You want to make sure they are in a position to choose the best courier to transport their consignment. For instance, a client may wish to make an order that will be delivered within 5-10 days, but by paying a few dollars more, the product will arrive within a shorter time than expected.

Participating in Farmers' Markets

A farmers' market is a regular and open market where farmers, their agents, or go-betweens sell foods they have produced. Farmers markets unite people to forge relationships and common interests that benefit farmers, consumers, and society.

Farmers get money, shoppers get the freshest and most delicious food available in their communities, and local economies benefit. Every farmer's market has its distinctive definition of the term 'local,' which is the region's agriculture, and this definition is conveyed to the public often. Farmers' markets also have policies and procedures that regulate their composition and operations. The farmers' market mainly comprises farms selling directly to the public food items that the farms have grown.

To safeguard farmers and consumers, some states have established official definitions that provide more detail about the market characteristics. Farmers' markets have rapidly expanded in the United States over the last decade.

It is not challenging to locate a farmers' market; here are some tips that can help. Farmers markets exist in all fifty states and can be found in all areas, including Main Street and city areas, parks, parking lots, sidewalks, and shopping malls.

To locate a market near you, try asking the people around you, your friends, Google, and your colleagues, or type it into the USDA's Farmers Market Directory or LocalHarvest.org and EatWellGuide.org. Many states also have a farmers' market association to help with the information.

Major regional market networks exist in some cities, such as Greenmarket in New York, Sustainable Food Center in Austin, FreshFarm Markets in Washington, D.C., and Neighborhood Farmers Market Alliance in Seattle. Most markets are single-site or multi-site entities that run independently or with the support of a city or a nonprofit organization, and some evolve into separate nonprofit markets as they expand.

Regardless of the farmers' market structure, there is always a market manager who polices the market bylaws and handles the day-to-day operations of the market. The market manager is usually the most appropriate person to contact.

Be sure that you qualify to participate in the study. Usually, individuals who cultivate crops and sell what they grow on their farm stand a chance to be awarded. There are a few requirements to qualify for each farmers' market.

If you meet all the requirements, apply for participation as a vendor. If reimbursement is required, provide a vendor agreement and a completed W-9 form for tax purposes. You will be assigned a seasonal vendor number before accepting vouchers or payments.

Spend most of the season operating in authorized farmers' markets or farmstands. Interact with consumers, promote your products, and develop relationships. However, note that there may be specific rules for each market that you need to find out from the market organizers. Good luck with your farmers' market business!

Chapter 12: Expanding Your Beekeeping Operation

Congrats on your thriving business. If you scale your business and go big, we are with you. Here's how to take your next steps.

Scaling Up: Adding More Hives

Before adding more colonies, you should always take stock of what you have. Think of the space available, the equipment that can be used, and the time that can be afforded. Expanding requires sufficient hives, frames, feeders, and protective wear. Knowing all the legal requirements that govern beekeeping in your area is essential. Consider the expenses involved and develop a clear strategy for acquiring the bees, hives, and other requirements for the venture.

To expand effectively, acquire brooks from healthy, productive bees from reliable sources. Consider the climate and conditions in your location to choose the suitable bee races. Transport the bees properly to avoid stressing them. Make enough space to accommodate new hives; ideally, they should be shaded with well-drained soil. Construct barriers, including fences, around the hives to shield them.

Proper construction of the hives and constant supervision of the bees is crucial in a good apiary. In terms of assembling the hive components and frames, adhere to the manufacturer's recommendations. They should ensure that the colonies are well supplied with feed and water, especially at the beginning of the colonies. Colonies should be visited for pest and disease control, defect correction, and checking the hives' strength and productivity.

Be observant and record the inspections, honey yield, and any other aspects that may affect the efficiency of the hives. This way, regular checks assist in detecting patterns and making the right decisions as per the system's improvement. Adopt appropriate technologies in bee farming, attend training and seminars, and be part of the beekeepers association to make you more knowledgeable.

To increase the number of colonies, consider sourcing nucleus colonies or packages. Nucs are small colonies with bees, brood, honey, and pollen; locally reared-queens perform better. Another method is splitting your current colonies through the simple divide or double-screen methods.

The primary method of swarm control is the divide method, in which one chooses a strong colony and divides the brood into two halves between two boxes. The queen is put into the lowest box or locked in a queen cage. If the queen is not seen, a queen excluder is used to search for her. The second box is then relocated, and a new queen is brought the next day of the week.

The double-screen method entails positioning a screen on top of the brood chamber. The queen and most broods are in the bottom box, and the upper box is queenless. A new one is implanted upon establishing that the colony has no queen.

The queen's health is essential to the colonies' productivity. Young queens are more productive and do not swarm as much as the other queens. Replace old queens with new ones and paint the new queen with different colors from the old one. Methods include candy cages, nucs, the newspaper method, push-in cages, virgin queens, or queen cells, which can guarantee a continuous supply of young queens, hence a productive business in beekeeping.

Diversifying Your Products

Expanding your beekeeping business will help increase revenue, productivity, and profitability. Besides honey, bees have other products that consumers prefer in the current society. For example, beeswax makes candles and cosmetics, while royal jelly is used for its nutritional value. There are also some related health benefits related to propolis and pollen. If you sell products you haven't diversified, you can sell them to other markets.

Second, it is possible to provide services concerning bees, although this choice is not as straightforward as the previous one. Crop pollination is vital to farmers, while bee control is vital to homeowners or business people. Besides generating revenues, these services also enable the public to appreciate the importance of bees in our environment.

This shows that the only way to grow is through partnership. Local businesses can be good customers for you or can help improve your business. The communities can also be educated, especially when you organize a conservation program or project that also markets your business and the importance of conserving bees.

Thus, increasing the capacity of your bee farming business implies that it is not solely about increasing honey production. You can get many other things from bees, services you can provide, and cooperation with other companies and societies to develop your business and support the stability of our environment.

Part 7:
Tips and Tricks

Chapter 13: Insider Tips for Successful Beekeeping

By now, you know everything you need to know about getting started with beekeeping and scaling this hobby to a profitable business. In this chapter, we are about to give you some insider tips that are highly instrumental in defining your success.

Managing Bee Health and Pests

Pest control and bee health are essential in bee farming. Although pesticides are helpful in agriculture, they are dangerous to bees, mainly when misapplied. Knowing the impacts of pesticides and the presence of residues is vital to protecting bees.

The toxicity of pesticides to bees is determined by the LD50, which is the lethal dose that would kill half of the test bees. A lower LD50 value means that the pesticide is more toxic, so its application should be carried out carefully so as not to endanger bees. Some pesticides are harmful initially but become less toxic as they are metabolized shortly after application. The RT25 value reflects the time needed for the pesticide residues to decrease to the level that results in 25% of the initial mortality rate of bees.

Regarding the timing of application, it is vital to remember that cooler temperatures somewhat hinder this detoxification process.

To reduce the effects of pesticides on bees, use the sprays when bees are not active, that is, during the night or when trees are not releasing pollen. This prevents the pesticide from drifting to the beehives and prevents bees from foraging on the contaminated plants. Do not use synthetic pesticides, fertilizers, or herbicides; instead, go for organic ones that are not harmful to bees. This involves planting trees that provide nectar and pollen, which help in bee reproduction and their general well-being.

It is helpful to give bees access to water; one could place a bee bath with clean water and pebbles for them to sit on. When the water supply is insufficient, we supplement the bees with food such as nectar and pollen to help maintain healthy hives. Thus, this approach to managing nutrition can significantly improve the health and productivity of bees.

Bee health and pest management are essential in sustainable beekeeping. Pesticides are toxic to bees by nature, so the steps of low toxicity, low residue pesticides, and bee-friendly farming are fundamental. Through these measures, beekeepers will be able to enhance the quality of their stock and, in so doing, support the quality of bees in general.

Enhancing Hive Productivity

Environmental and biological factors contribute to variations in honey yield. The weather conditions and the availability of pollen and pests in the area more or less determine honey yield. Temperature, humidity, rain, and the rest are the main climatic factors affecting bees. Adverse weather factors like scorching and dry weather or very cold and wet weather may slow down the movement of bees to look for nectar and water, which is vital for honey production.

The honeybees are interested in the nectar since they need the pollen to synthesize protein to feed on their young ones. This, in turn, has the impact of slowing down the colony's development and, in the case of honey bees, less honey production. Also, there is a danger in the lack of flowers in the surrounding environment because bees may struggle to find the nectar for honey production. Pest, especially parasites like mites, might result in compromises or even death of the bees, hence, weakness of the colony and low honey production.

Natural conditions that influence honey yield include diseases and parasites, which are biological factors that affect honey yield. Bee diseases like Nosema and American Foulbrood can infect the bees and may harm the entire colony or even kill it. Such information is valuable as it assists beekeepers in finding strategies for maintaining the colony and increasing honey production.

Beekeepers need to follow certain principles to understand how to get the best bees and, thus, the highest honey production. The portions should be sufficient to cater to the food needs of the people, and this is more so where food and water are scarce. The same is true for other nutrients because feeding bees with other nutrients also boost honey production.

Here are some of the most crucial factors bees use in their honey-making process, considered some of the best-kept secrets. Proper beehive management includes ventilation and insulation because they assist in creating a suitable climate for bees to thrive and produce at the desired rate. Through practice in the way described above, the beekeepers will be able to have healthy bees that will result in a good honey yield.

Building a Beekeeping Community

Regardless of your level of experience, it is recommended that you join a beekeeping association because the group can be a source of information, friends, and a support system. Community involvement can be advantageous in many ways, from beginners to those who have been in the practice for many years.

Members of beekeeping organizations have the added advantage of networking. This allows beekeepers to communicate with others in the sector, find links to local services, and learn about current practices. These relations can benefit personal and professional growth by allowing us to gain confidence, ask questions, and learn from beekeepers with more experience in this business.

The community also has its advantages over individual living in the practical aspect. Beekeepers can learn the rules and regulations in the area, visit bee farms, learn about the suppliers and customers of bees and bee products, learn about the beekeepers' insurance, and exchange materials and equipment. Besides, it is an added advantage to be in an organization in beekeeping because one can get updated information on beekeeping, consultation from other experienced beekeepers and neighboring beekeepers, and even special offers on supplies and equipment for beekeeping.

Thus, beekeeping organizations play a role in developing opportunities for beekeepers. They inform society, establish new contacts via meetings, offer inexpensive tools and equipment, spread knowledge about beekeeping and bee products, and introduce new initiatives in this field.

Local beekeeping clubs are beneficial in that members can learn from other beekeepers and pass on some lessons to others in the same region. They can also help members comprehend various problems that affect bees and possible solutions to address those problems.

Other related activities that people may embark on include bee master classes or a beekeeping business, which may also be rewarding. Problems with technology in beekeeping, such as hive scales and sensors, can also enhance beekeeping and productivity.

Chapter 14: Troubleshooting Common Problems

The last chapter of this book will share some vital information regarding how to troubleshoot common problems you can encounter in beekeeping. Let's dive in!

Wax Moth Infestation

Identifying Wax Moth Infestation

A beekeeper should look for the following indicators to identify wax moth infestation in a beehive: webbing and tunnels, which are characteristic features of the honeycomb and have been left behind by the moth larvae. These tunnels are usually interconnected, referred to as having a spider-like web pattern, and can be located in any comb area.

When in the unguarded comb, the larvae continue consuming the wax until it is depleted. Another sign is the formation of patterns of bald brood due to bees uncapping some cells. These signs can help beekeepers determine the presence of wax moths and take necessary action against them.

Treating Wax Moth Infestation

Here's a step-by-step approach to treating wax moth infestation based on current best practices:

Freezing Infested Materials: Removing affected frames or combs from the hive and freezing them for at least 48 hours can kill wax moth eggs, larvae, and adults. This is an effective and non-toxic method of halting wax moths' life cycle.

Exposure to Sunlight: Placing infested frames or combs in direct sunlight can also help eliminate wax moth larvae and eggs. The heat and UV rays harm the larvae and help reduce infestation.

Using Moth Traps: Set up pheromone-based moth traps inside and around your beehive. These traps attract and capture adult moths, thereby reducing their population and preventing reproduction. You can also create DIY traps using a mixture of sugar, water, vinegar, and banana skin, which effectively attracts moths.

Essential Oils: Applying diluted essential oils like thyme, lemongrass, or tea tree oil around the hive can be a natural deterrent for wax moths. These oils have insect-repellent properties that help in protecting your hive from future infestations.

Chemical Treatments: In severe cases where natural and physical methods are insufficient, chemical treatments like Para Dichlorobenzene (PDB) and Bacillus thuringiensis (Bt) can be used. PDB is effective for treating stored combs, while Bt targets moth larvae. However, these chemicals should be used cautiously and per manufacturer instructions to avoid harming the bees.

Maintaining Hive Hygiene: Regularly cleaning the hive and removing old combs can prevent wax moth infestations. To discourage moth activity, ensure your beehives are well-ventilated and placed in well-lit areas.

Hive Beetles

Identifying Hive Beetles

Small hive beetles are another major menace that beekeepers are advised to look out for. The best way to determine their presence is probably by actually seeing them. These beetles are not fond of direct sunlight and will scamper away when exposed to light. They should not be confused with other insects that may inhabit the colony and have different appearances, such as earwigs. During an inspection, beekeepers should expect to see small hive beetles and be ready to act if they see them in the hive.

Treating Hive Beetle Infestation

There are several ways that you can employ to protect your bees from hive beetles.

Mechanical Traps: Oil tray traps under the hive catch larvae and adults by trapping them in oil as they fall. These trays fit under the hive and replace the bottom board, ensuring beetles can't escape once trapped. Swiffer pads or unscented dryer sheets placed on top bars inside the hive can also effectively capture adult beetles, as the beetles get stuck in the fibers.

Chemical Treatments: Chemicals like Checkmite+ are effective against both larvae and adult beetles. This treatment involves placing a strip inside the hive for about 42-45 days. It's crucial not to use this treatment during honey harvest and to wait 14 days after removing the strip before adding any honey supers. Permethrin can also be used to treat the soil around hives to kill beetle pupae, ensuring thorough ground treatment by removing vegetation and applying directly to the soil.

Diatomaceous Earth: Apply diatomaceous earth around the hive to kill larvae and pupae. This natural method involves dusting the ground thoroughly around the hives and can be combined with watering to enhance its penetration into the soil.

Boric Acid: Use boric acid applied to correx boards placed within the hive. Beetles are attracted to shortening on the boards, and then they consume boric acid, which is lethal to them. This method targets beetles inside the hive without significant harm to the bees.

Freezing Infested Frames: If beetle larvae are present in the frames, freezing them for 24 hours effectively kills all stages of beetles. This method is non-chemical and highly effective at stopping infestations from spreading .

Natural Predation: Employ chickens, if possible, as they can help by eating larvae and pupae around the hive. Chickens do not disturb the bees but help control the beetle population by consuming the larvae found on the ground near the hives.

Varroa Mites

Identifying Varroa Mites

Controlling Varroa infestation is essential for healthy bee colonies to be observed. Beekeepers should consider requiring mite-resistant stock in a bid to reduce the reproduction rate of Varroa. Several techniques can be applied to identify Varroa mites: sticky boards for obtaining the mite number, alcohol wash, or the most common method—sugar/ether roll. To counter this Varroa mite, beekeepers are encouraged to conduct frequent check-ups and treat their bees to eliminate this disease-causing parasite.

Treating Varroa Mites

Here's a detailed step-by-step guide on treating Varroa mites in bee colonies, pulling together up-to-date methods and practices:

Drone Brood Removal: This method targets the drone brood, which Varroa mites prefer for reproduction. Remove the drone frames just before the brood caps, interrupting the mite reproduction cycle.

Powdered Sugar Dusting: Dusting bees with powdered sugar encourages grooming behavior, causing mites to fall off the bees. This method is effective when combined with other strategies and should be repeated regularly for best results.

Screened Bottom Boards: Using screened bottom boards allows mites to fall out of the colony, preventing them from reattaching to the bees. This method is more effective when used in combination with other treatments.

Soft Chemicals: Treatments like thymol, formic acid, and oxalic acid are preferred. Thymol, extracted from thyme, is effective but cannot penetrate brood cells, so it's better used when the mite load is primarily on adult bees. It's most effective at moderate temperatures (between 68°F and 77°F). Formic acid can penetrate wax cappings and effectively target mites in brood cells, but it's sensitive to high temperatures, which can endanger the queen and brood.

Oxalic Acid: Useful during brood less periods or combined with other methods, as it doesn't affect mites in capped cells.

Regular Monitoring and Integrated Pest Management (IPM): Continuously monitor mite loads using sticky boards or alcohol washes to determine the infestation level and effectiveness of your treatments. Implementing IPM involves combining various methods and adjusting them based on effectiveness and environmental impact.

Treatments should be carefully timed to match the colony's lifecycle and mite reproduction cycles. For instance, applying oxalic acid when the colony is broodless maximizes impact, as all mites will be on adult bees, making them more vulnerable. There are also some alternative methods to this approach.

Nosema Disease

Identifying Nosema Disease

However, bee colonies are vulnerable to other diseases, including Nosema. Thus, to diagnose Nosema, the beekeeper can use live or recently dead bees taken from the entrance or the top bars of frames. These bees can then be taken to the laboratory, where they can be put under the microscope to count the spores in the honey bee gut.

If the spore count in the colonies is above 1 million spores per bee, they may experience dwindling, but this is not necessarily the case. Nosema is a disease that beekeepers should consider due to its ability to infect a colony and cause significant problems, which is why it is important to monitor it and treat it as soon as possible.

Treating Nosema Disease

The most effective methods for treating nosema are as follows.

Fumagillin Treatment: Fumagillin is the primary chemical used to control Nosema. It's mixed with sugar syrup and fed to bees to reduce spore counts. However, it's essential to note that Fumagillin can have residues in honey, and its use is regulated in some regions due to potential resistance issues.

Probiotics: Adding probiotics to the bees' diet helps maintain healthy gut flora, which is crucial in fighting off infections like Nosema. Probiotics can enhance the bees' immune response and overall health.

Essential Oils: Some beekeepers use thyme and lemongrass oils, which have shown potential in supporting bee health and combating Nosema. These should be used carefully to prevent any adverse effects on the bees.

Combination Therapy: Combining different treatment methods, such as antibiotics and probiotics, might provide a more comprehensive approach to managing Nosema, helping to reduce the spore load while supporting the bees' overall health .

Preventative Measures

Preventive measures are crucial in apiculture to have healthy colonies yielding good honey. Some practices to avoid problems such as pest infestation, diseases, swarming, and general management of the hives include. Some of the practices to prevent issues such as pest infestation, diseases, swarming, and general management of the hives include:

Measures that can be taken to combat pests like tracheal mites include increasing the bee stock and breeding for strains resistant to the pests. Other treatments like menthol or formic acid can also be used. Another method of colonial inspection is to inspect for the presence of mite infestation, which is helpful in the early detection of the disease.

To prevent ants and other pests from invading the hives, stands can be placed under them, traps placed, or a layer of glue placed on the legs of the hives. It is also equally important to note that the area surrounding the hives should be clean, as this will help eliminate ants.

For diseases such as the American Foulbrood (AFB), the brood frames must be inspected, and those affected by the disease must be identified. In this case, an antibiotic such as oxytetracycline has to be used. Sanitary conditions of hives and avoiding using equipment that has been in contact with infected ones are also adequate. To prevent EFB, one should take samples as indicators and administer oxytetracycline if needed, feed the colonies healthily, and ensure sufficient aeration. The measures that can be taken to control the disease include selecting and using chalkbrood-resistant bee stocks, proper management that provides for proper ventilation and low humidity of the hives, and removing and replacing the infected brood comb with a new one.

Some strategies that may be useful in controlling swarming include creating supers to reduce overcrowding, searching for Queen cells and destroying them if found, splitting solid colonies, and others. The other ways to prevent swarming include controlling the queens by ensuring that an old queen is replaced with a new one annually, utilizing the marked queens that will help identify their age, and ensuring the availability of adequate space for brood in the colonies.

As for productivity, it is possible to place the hives in an area with good exposure to sunlight, protected from winds, and close to the forage. Another factor that can enhance productivity is to ensure that the hives are always provided with clean water and supplement feeding when the natural resources are rare. Some of the measures that should be practiced to improve the health and productivity of the colonies include sanitization of the hives, removal of old and weak comb, proper provision of the hives with adequate ventilation, proper rotation and replacement of the brood frames as well as proper observation of the patterns of the brood.

Conclusion

(Keep Reading, There's a Bonus for You at the End!)

Reflecting on Our Beekeeping Journey

Now, you are at the last point of the Ultimate Beekeeping Guide, so it is time to recall the fantastic journey you have made at the beginning. This book is designed to lead the reader through the bee farming process in a systematic manner, which is thrilling and very lucrative. To the first-time beekeeper or even the professional beekeeper who needs to sharpen his knowledge, we have given you helpful information, knowledge, and confidence in your beekeeping endeavors.

Our journey started in Part 1, where we provided a brief introduction to beekeeping and the role of bees in our ecosystems. Acquiring knowledge of bees' activity in pollination and the significance of species' conservation was the foundation for understanding the importance of beekeeping for preserving the environment and agriculture. We also explained how to employ this guide, which gave the reader an orientation in the variety of the materials provided in this book.

In part 2, the history of beekeeping was discussed, beginning from the early ages to the present practice of beekeeping. This section describes how bees help make honey together with beeswax and how they help increase the yield of gardens through pollination. We first ensure that you are familiar with the fundamental terms used in apiculture and then go further to assist you in choosing the first bee colony. The type of bees and their behavior, source of acquiring your bees, and legal matters are some of the first things to consider when practicing bee farming.

In Chapter 4, the major emphasis was on the Langstroth, Top Bar, and Warre hives. We described the strengths and weaknesses of each type and made a conclusion to help you make the choice based on your needs. For those who are determined to engage in practical activities, we also gave procedures on how to build your own beehive, including the list of things required, how to build the beehive, and how to introduce the bees.

As your hive is ready, chapter 5 provides a basic idea about installing and maintaining your bee colony. We described how to get the bees into the new hive and what should happen in the first days to weeks after transferring the bees. After that, in the subsequent chapters, we realized why the management of hives is seasonal and what to do in each of the seasons, namely spring, summer, autumn, and winter. It was also advised that you should maintain the hives to observe the bees and address any issue that may be present early enough so that it does not accumulate serious complications.

In part 5, you were taken through what beekeeping offers you, including honey and beeswax. We talked about when to harvest honey, how to remove it from the comb, methods of processing honey, and the proper techniques for storing and packaging your honey. Furthermore, we have learned that there are many uses for beeswax.

For those interested in turning their passion into a business, Part 6 helped us grasp how honey and beeswax can be marketed. We discussed marketing strategies, developing an online selling platform, and selling at farmer's markets. The only way to continue your beekeeping business is to expand it, and this can be a giant leap. That is why this section provides more complex techniques and tips on increasing your beekeeping business.

Part 7 synthesizes information about the company's functions and strategies for possible issues. We discussed the proper methods for managing diseases and pests in bees, how to enhance bee breeding, and how to foster a good beekeeper society. Solutions for possible difficulties and additional measures to protect yourself were provided to prepare you for any problems that might occur.

Before putting this book to rest, it is crucial to acknowledge that beekeeping is a dynamic and evolving discipline. Every hive and beekeeper's journey is unique, so it will be different. This guide has given you the fundamentals and essential pointers on the beekeeping business, but the learning does not stop here. Study your bees, introduce yourself to as many people as possible, and observe your bees. They are your best teachers.

The Future of Beekeeping

In the future, beekeeping is expected to have a better future for the following reasons. The findings in research and technology are constantly improving knowledge about bees and their need to enhance bee farming. Technology is slowly seeping into the beekeeping process through the emergence of solutions to conventional methods. IoT and other manufacturing technologies have led to new beekeeping approaches that tech startups are embracing for increased challenges such as habitat loss, colony collapse disorder, and commercial beekeeping. IoT and app-based monitoring, drones and robots for pollination, and similar practices are examples of sustainable practices. Such enhancements are aimed at helping beekeepers adequately manage the internal conditions of hives and honey production.

The following are some ways that availing technology has been of great help to beekeepers in the modern world. It assists in feeding data intelligence on hive health and activity that can identify issues such as pest invasions and nutrient deficiency. There are such technologies as traps with semiochemicals to detect the presence of varroa mites and drones to control the colonies' state and the bees' efficiency. AI is also used to work out better conditions for bees through hives, climate, and other factors, as well as the ratio of bees' activity and flowers containing nectar.

New digitally assisted methods may appear for the control of pests and diseases in apiculture. Currently, tracking systems enable constant monitoring of pests, which helps prevent pest damage and maintain record-keeping. Using data in decision-making through smartphones and wearable technology can easily predict the impacts of pests or diseases based on different factors. Advanced techniques of watering through intelligent watering systems and the platforms of precision agriculture also increase the effectiveness of pest control.

Negligence in proper colony management is one of the primary causes of poor health among hives, and the use of basic yet efficient tools and methods is the key to proper health management of the colonies. Letting the public know the importance of bee farming can go a long way in raising the number of people who practice the activity and improving the quality of bee farming. Feeding bees through natural diets and feed supplements can improve the productivity of the hives and also the quality of the honey produced. In addition, there is the development of proper strategies for bee farming that will not be a nuisance to the environment and, at the same time, help the bees to cope with the effects of climate change. Public awareness campaigns and research collaborations are also helping beekeeping as a sustainable practice so that healthy bees are conserved for future generations.

Bonus Section

Thank you for reading the Ultimate Beekeeping Guide to the very end! To show our appreciation for your hard work and loyalty, we have provided a special bonus section, which will give you some extra tips to help you on your beekeeping adventure. These resources are created to help you be more organized, keep healthy hives, and get the most out of beekeeping.

Beekeeping Calendar

First, let us start with the Beekeeping Calendar. This calendar is a unique reference that will help you understand the yearly activities essential for the proper functioning of the bees and the hives. Every season poses some challenges and some opportunities, and it is up to you to counter-check these challenges and grab these opportunities.

Spring

Your bees will be more active during warmer months of the year as winter dissipates and spring takes over. Spring is the season to check your hives to determine whether they are alive and to check their health status. Make sure the queen is still in the hive and has a proper brood pattern. Look for food stores, and if there are none, consider feeding sugar syrup if available. Clean and repair equipment, as well as add new supers to accommodate the increasing population of bees.

Summer

Summer is the period of the heaviest activity of your bees. Supervise for swarming behavior and act to prevent further occurrence. It is also important to carry out routine inspections of the hives for pests and diseases, including varroa mites, and take the appropriate action. Ensure that the hives are well-ventilated to avoid heat buildup, and harvest honey that has begun to crystallize during the early part of the year. Make sure that the bees are left with enough honey.

Fall

In the fall, make your last honey harvest and then block hive entrances to prevent robbing. Spray for pests to make sure your bees are healthy as they head into winter. Ensure that you check whether the hives have a stock of foods that can sustain them during the winter season, as this is very important.

Winter

Winter is not an active season for bees. However, it is necessary to visit hives periodically to check the ventilation and humidity levels. Ensure the hives are covered to protect them from wind and predators, check food supplies, and replenish them if required. Correct wintering will allow colonies to appear on the stage in the spring.

Month	Key Tasks
January	Check food stores; add candy boards if necessary.
February	Prepare for spring; clean and repair equipment.
March	Begin inspections; check for queen health and brood viability.
April	Add supers as needed for nectar flow; monitor for swarming.
May	Increase swarm prevention measures; manage splits if necessary.
June	Monitor honey flow; ensure adequate super space for honey.
July	Begin mid-season honey harvest; treat for varroa mites if needed.
August	Prepare hives for the end of honey flow; start winter preparations.
September	Feed bees if necessary; continue winter preparations.
October	Secure hives from cold; finish winter preparations.
November	Check hives for food stores; wrap hives in colder regions.
December	Minimal inspections; plan for the next year.

Hive Inspection Checklist

It is important that hives are checked on a routine basis as this will help assess the health of the bees as well as the amount of honey produced. The checklist is helpful in making sure that there is no aspect left behind every time the inspectors go through the facilities. Here's an example of what a thorough checklist might include. Here's an example of what a thorough checklist might include:

- **Queen Activity**: Check on the queen, see that she is present, and determine her laying pattern. The brood pattern of a queen should be uniform and packed throughout her lifetime without any gaps.
- **Brood Pattern**: Check for regularity in the distribution of brood since patchy distribution may be a result of the queen's problems or diseases.
- **Food Stores**: Determine the amount of honey and pollen and check that there is enough, especially before the onset of the winter season.
- **Signs of Disease or Pests**: Search for signs of other familiar diseases such as the American foulbrood or pests such as varroa mites. This is why diagnosis in the early stages is important to help a patient receive the right treatment.

- **Hive Structure**: Ensure that there is no formation of cracks or openings in the hive that would allow pests or weather to penetrate into the hive. Also, check for adequate ventilation and the state of insulation and weatherproofing.

To help you get started, we have prepared checklists for your beekeeping inspections in the next few pages. You can use this checklist when you get started with your beekeeping journey, and then you can buy or create your own checklist when you have somewhat settled into it. These additional tools are designed to help you with the main parts of the book and to help you sustain a successful apiary. This is the reason why beekeepers need to inspect their bees frequently and always be prepared for seasonal changes.

Once again, thank you for choosing the Ultimate Beekeeping Guide. We trust that these resources will help improve your beekeeping experience and continue to yield positive results. Blessed beekeeping, and may your hives be healthy, productive, and alive!

Checklist 1

*Instruction: Put (✓) when the answer is Yes and put (X) when the answer is No

Inspection Date/Time:		Hive #1	Hive #2	Hive #3	Side Notes
General Observations	Are bees active?	☐	☐	☐	
	Pollen Entry Observed?	☐	☐	☐	
	Any robbing signs?	☐	☐	☐	
Brood Check	Uniform brood pattern?	☐	☐	☐	
	Larvae appear healthy?	☐	☐	☐	
	Royal jelly in cells?	☐	☐	☐	
Pest Assessment	Varroa mites present?	☐	☐	☐	
	Ants detected?	☐	☐	☐	
	Wax moths' signs?	☐	☐	☐	
Hive Capacity	Frames mostly occupied?	☐	☐	☐	
	Adequate nectar storage?	☐	☐	☐	

Ultimate Beekeeping Guide

Weather Conditions	Suitable for bees?	☐	☐	☐	
	Are weather issues noted?	☐	☐	☐	
Overall Hive Health	Hive stability observed?	☐	☐	☐	

Notes

_____
_____
_____
_____
_____
_____
_____
_____

To-Do List:

Checklist 2

*Instruction: Put (✓) when the answer is Yes and put (X) when the answer is No

Inspection Date/Time:		Hive #1	Hive #2	Hive #3	Side Notes
General Observations	Are bees active?	☐	☐	☐	
	Pollen Entry Observed?	☐	☐	☐	
	Any robbing signs?	☐	☐	☐	
Brood Check	Uniform brood pattern?	☐	☐	☐	
	Larvae appear healthy?	☐	☐	☐	
	Royal jelly in cells?	☐	☐	☐	
Pest Assessment	Varroa mites present?	☐	☐	☐	
	Ants detected?	☐	☐	☐	
	Wax moths' signs?	☐	☐	☐	
Hive Capacity	Frames mostly occupied?	☐	☐	☐	
	Adequate nectar storage?	☐	☐	☐	

Weather Conditions	Suitable for bees?	☐	☐	☐	
	Are weather issues noted?	☐	☐	☐	
Overall Hive Health	Hive stability observed?	☐	☐	☐	

Notes

_____
_____
_____
_____
_____
_____
_____
_____
_____

To-Do List:

Checklist 3

*Instruction: Put (✓) when the answer is Yes and put (X) when the answer is No

Inspection Date/Time:		Hive #1	Hive #2	Hive #3	Side Notes
General Observations	Are bees active?	☐	☐	☐	
	Pollen Entry Observed?	☐	☐	☐	
	Any robbing signs?	☐	☐	☐	
Brood Check	Uniform brood pattern?	☐	☐	☐	
	Larvae appear healthy?	☐	☐	☐	
	Royal jelly in cells?	☐	☐	☐	
Pest Assessment	Varroa mites present?	☐	☐	☐	
	Ants detected?	☐	☐	☐	
	Wax moths' signs?	☐	☐	☐	
Hive Capacity	Frames mostly occupied?	☐	☐	☐	
	Adequate nectar storage?	☐	☐	☐	

Weather Conditions	Suitable for bees?	☐	☐	☐	
	Are weather issues noted?	☐	☐	☐	
Overall Hive Health	Hive stability observed?	☐	☐	☐	

Notes

_____
_____
_____
_____
_____
_____
_____
_____

To-Do List:

Checklist 4

*Instruction: Put (✓) when the answer is Yes and put (X) when the answer is No

Inspection Date/Time:		Hive #1	Hive #2	Hive #3	Side Notes
General Observations	Are bees active?	☐	☐	☐	
	Pollen Entry Observed?	☐	☐	☐	
	Any robbing signs?	☐	☐	☐	
Brood Check	Uniform brood pattern?	☐	☐	☐	
	Larvae appear healthy?	☐	☐	☐	
	Royal jelly in cells?	☐	☐	☐	
Pest Assessment	Varroa mites present?	☐	☐	☐	
	Ants detected?	☐	☐	☐	
	Wax moths' signs?	☐	☐	☐	
Hive Capacity	Frames mostly occupied?	☐	☐	☐	
	Adequate nectar storage?	☐	☐	☐	

Weather Conditions	Suitable for bees?	☐	☐	☐	
	Are weather issues noted?	☐	☐	☐	
Overall Hive Health	Hive stability observed?	☐	☐	☐	

Notes

_____
_____
_____
_____
_____
_____
_____
_____
_____

To-Do List:

Checklist 5

*Instruction: Put (✓) when the answer is Yes and put (X) when the answer is No

Inspection Date/Time:		Hive #1	Hive #2	Hive #3	Side Notes
General Observations	Are bees active?	☐	☐	☐	
	Pollen Entry Observed?	☐	☐	☐	
	Any robbing signs?	☐	☐	☐	
Brood Check	Uniform brood pattern?	☐	☐	☐	
	Larvae appear healthy?	☐	☐	☐	
	Royal jelly in cells?	☐	☐	☐	
Pest Assessment	Varroa mites present?	☐	☐	☐	
	Ants detected?	☐	☐	☐	
	Wax moths' signs?	☐	☐	☐	
Hive Capacity	Frames mostly occupied?	☐	☐	☐	
	Adequate nectar storage?	☐	☐	☐	

Ultimate Beekeeping Guide

Weather Conditions	Suitable for bees?	☐	☐	☐	
	Are weather issues noted?	☐	☐	☐	
Overall Hive Health	Hive stability observed?	☐	☐	☐	

Notes

_____
_____
_____
_____
_____
_____
_____
_____
_____

To-Do List:

Checklist 6

*Instruction: Put (✓) when the answer is Yes and put (X) when the answer is No

Inspection Date/Time:		Hive #1	Hive #2	Hive #3	Side Notes
General Observations	Are bees active?	☐	☐	☐	
	Pollen Entry Observed?	☐	☐	☐	
	Any robbing signs?	☐	☐	☐	
Brood Check	Uniform brood pattern?	☐	☐	☐	
	Larvae appear healthy?	☐	☐	☐	
	Royal jelly in cells?	☐	☐	☐	
Pest Assessment	Varroa mites present?	☐	☐	☐	
	Ants detected?	☐	☐	☐	
	Wax moths' signs?	☐	☐	☐	
Hive Capacity	Frames mostly occupied?	☐	☐	☐	
	Adequate nectar storage?	☐	☐	☐	

Ultimate Beekeeping Guide

Weather Conditions	Suitable for bees?	☐	☐	☐	
	Are weather issues noted?	☐	☐	☐	
Overall Hive Health	Hive stability observed?	☐	☐	☐	

Notes

_____
_____
_____
_____
_____
_____
_____
_____
_____

To-Do List:

References

[1] Kaiser, C. (2022, September 1). *Beekeeping and Honey Production.*
https://www.uky.edu/ccd/sites/www.uky.edu.ccd/files/honey.pdf

[2] Barnes-Jewish Hospital (n.d.). *BEE VENOM USED TO KILL HIV.*
https://www.barnesjewish.org/Newsroom/Publications/Innovate/Fall-2013/Bee-Venom-Used-to-Kill-HIV

[3] Clemson University (n.d.). *Rules and Regulations.* Clemson Cooperative Extension.
https://www.clemson.edu/extension/pollinators/apiculture/rules-regs.html

[4] Michigan State University (n.d.). Michigan Beekeeping Rules and Regulations. MSU Extension Pollinators & Pollination.
https://www.canr.msu.edu/resources/starting_and_keeping_bees_in_michigan_rules_and_regulations

[5] (n.d.). *The Honey (England) Regulations 2015*. Legislation.gov.uk.
https://www.legislation.gov.uk/uksi/2015/1348/schedules

[6] (n.d.). *Animal Health Act BEE REGULATION*. BC Laws.
https://www.bclaws.gov.bc.ca/civix/document/id/complete/statreg/3_2015

www.ingramcontent.com/pod-product-compliance
Lightning Source LLC
LaVergne TN
LVHW010307070426
835512LV00029B/3498